Wânia Olívia Costa

The city invades the forest

Wânia Olívia Costa

The city invades the forest

Environmental impacts in the Tijuca Forest - Rio de Janeiro - Brazil

ScienciaScripts

Imprint

Any brand names and product names mentioned in this book are subject to trademark, brand or patent protection and are trademarks or registered trademarks of their respective holders. The use of brand names, product names, common names, trade names, product descriptions etc. even without a particular marking in this work is in no way to be construed to mean that such names may be regarded as unrestricted in respect of trademark and brand protection legislation and could thus be used by anyone.

Cover image: www.ingimage.com

This book is a translation from the original published under ISBN 978-613-9-66635-5.

Publisher:
Sciencia Scripts
is a trademark of
Dodo Books Indian Ocean Ltd. and OmniScriptum S.R.L publishing group

120 High Road, East Finchley, London, N2 9ED, United Kingdom
Str. Armeneasca 28/1, office 1, Chisinau MD-2012, Republic of Moldova, Europe
Printed at: see last page
ISBN: 978-620-8-05229-4

I dedicate this work to my great source of inspiration, my daughter Júlia. To my parents Maria and Domício (in memorian), who always taught me that study is the main tool for growth and independence for women. And the greatest lesson of all: to have respect, love and care for all living beings: animals, plants, women and men.

ACKNOWLEDGEMENTS

I would like to thank the IFRJ (Federal Institute of Education, Science and Technology - Rio de Janeiro), my dear supervisor Ana Paula Silva, all the teachers and friends. To PARNA-Tijuca (Tijuca National Park), for the willingness of the staff and volunteers to carry out this research. Especially to my friend Marcelo Lopez.

To my editor André Lemos.

To my Mum, my daughter, my family and friends. To my editor, André Lemos. To my therapist and friend Marscyelle Brasil.

"The basic laws of nature have not been repealed, only their features and quantitative relationships have changed, as the world's human population and its prodigious energy consumption have increased the possibility of altering the environment. As a result, our survival depends on knowledge and intelligent action to preserve and improve environmental quality through harmonious and non-damaging technology."

Eugene P. Odum (1983)

SUMMARY

CHAPTER 1

HUMAN BEINGS AND NATURE:

We human beings have a relationship of direct dependence on nature and its natural resources. This statement highlights a basic biological issue. However, anthropogenic actions that have left negative traces for centuries present us with a paradox. In modern society, the general trend is towards a dissociated vision between human beings and nature. The result is the great environmental crisis we see today. This view is the result of the strong anthropocentrism of today's societies. In contrast, most primitive societies established a relationship of interaction with the environment. In the case of Brazil, this relationship can be exemplified by the indigenous groups and, later, the Africans. It wasn't unbridled exploitation, as the Europeans practised when they colonised Brazil. In fact, there was interaction, as they saw themselves as part of the natural systems. The way the land was managed was sustainable. They used crop rotation and took from the land only what was necessary for subsistence (JECUPÉ, 1998).

According to Drummond (1998), Brazilian Amerindians would fit the bill of "mythical" in terms of their religious beliefs and their appreciation of natural elements. For them, fauna, flora, land, rivers, forests and mountains belong simultaneously to the sacred world and the world of the profane. In the same way, the landscape is made up of natural elements that are positively valued and codified in legends, rituals and entities. It is "sacred" or, at the very least, contains "sacred" elements (DRUMMOND, 1988).

An interesting aspect of the relationship between worldviews and primitive societies is analysed by researcher Sahtouris (1991). In her studies, the author differentiates between agricultural societies and nomadic hunting societies. She considers that agricultural societies were well planned and managed; there were large cities and agricultural technology at the same time. These groups were egalitarian, peaceful and democratically advanced societies. In contrast, nomadic hunter-gatherer societies were made up of invaders and conquerors, experienced in the use of weapons. These peoples were not egalitarian, establishing themselves competitively due to the harsh environment.

Agricultural societies considered nature to be a great mother, alive and changing; people were part of this being (SAHTOURIS, 1991). In contrast, for nomadic hunting societies, nature was separate from both gods and people. Nature was created by a god outside of it; it was a gift to be used and exploited. Men and their gods enjoyed an external and superior position to nature.

We can therefore see that the separation between society and nature is quite old. It is consumer society, however, that has made this separation more intense, leading to what can be described as a rupture. When profit and wealth become the main objective of social relations, natural resources become the means by which wealth is generated.

With regard to Brazilian society, more specifically, we can say that, ever since the accounts of our colonisation, the abundance of natural resources has prevented the formation of a feasible idea of a "limit of use" on the part of the explorers (HEYNEMANN, 1993)[1] . Nowadays, however, there is a consensus on the need to balance economic development and environmental preservation as premises for sustainable development - a term defined in the Stockholm Declaration in 1972.

As for the country's conservation policy, while on the one hand there is already a clear interest in environmental preservation and protection, on the other hand, these objectives still need to be satisfactorily articulated with economic and social demands. Both aspects are important for maintaining the nation's sustainable growth. Individuals and organisations must therefore opt for a conservationist approach.

Sometimes, however, these aspects clash. The very least that is expected of a company today is that it acts in accordance with current legislation. Today, important legal actions can already be found in a wide variety of areas. Even if the interest of

---

[1] During Brazil's colonial and imperial periods, there are records of initiatives, including legal ones, aimed at protecting forests. However, these initiatives were constantly vetoed and never implemented (HEYNEMANN, 1993). It is important to highlight them, however, in order to understand more clearly the dichotomy of conservation versus exploitation. Furthermore, these initiatives are the beginnings of environmental legislation in Brazil, signalling its progress along the way.

large companies is to gain good visibility in the market through green marketing, concrete and relevant actions need to be taken. Environmental treatment cannot be restricted to words.

Environmental interest is usually expressed in actions by organised social groups (local, national and international) such as Greenpeace, the SOS Mata Atlântica association and government initiatives around the world. But it must be expanded. The threat of scarcity of natural resources affects the economy as a whole and the maintenance of living beings on the planet.

Brazilian legal provisions today impose reformulations on people's activities and especially on enterprises that cause environmental impact, in which they are obliged to carry out the EIA/RIMA (Environmental Impact Study/Environmental Impact Report), as a pre-requisite for environmental licensing.[2]

In conclusion, we cannot limit ourselves to isolated actions or merely reactive movements. The commitment must be global. Nature conservation and economic development will only be effective if the concept of sustainable development is implemented alongside economic activities. It is important to understand some of the perspectives related to this binomial. Let's now look at a very important topic concerning Conservation Units.

CONSERVATION UNITS

The creation of protected areas has become an imperative for humanity. For some time now, an extensive conservation policy has been established around the world. Since ancient times, however, we can already find the practice of earmarking areas for preservation, whether for natural, social, cultural, religious reasons (sacred groves) or for social status. One example is the royal hunting parks

---

(HEYNEMANN, 1993), such as Hyde Park in London, England, which was restricted to hunting until the reign of the English king James I, when access to the park became limited. It wasn't until 1637, under the reign of the English king Charles I, that the public was allowed in (Hyde Park -Senior Playground Feasibility Study, 2012).

The establishment of these areas arose from intense processes of occupation and appropriation of the land and its natural resources. At various times, human society has realised the importance of creating protected natural areas, with the aim of improving quality of life and generating environmental services (THOMAS E FOLETO, 2012).

In most contemporary countries, the practice of environmental protection for specially protected natural spaces already exists. The aim is to guarantee the protection of biodiversity and the sustainable use of resources. The delimitation of certain portions of the territory and limiting the use of their resources has become an important strategy for protecting the environment. Furthermore, as they produce spaces with specific dynamics and differentiated administration, the creation of these areas is considered an important territorial planning strategy managed by the state (MEDEIROS, 2006; MEDEIROS and YOUNG, 2011).

The IUCN, The World Conservation Union, has conceptualised *Protected Areas* as "an area of land or sea specially dedicated to the protection and preservation of biological diversity, as well as associated natural and cultural resources, and managed through legal or other effective means" (SCHERL, 2006). It is thus a geographically defined area where the main objective is the *in situ* protection of environmental attributes.

It is important to emphasise that Protected Areas are not legally protected "by chance", because they are environmentally important. Their structure, dynamics and function contribute to maintaining people's quality of life. They therefore need specific regulations to guarantee protection and prevent degradation.

According to Diegues (2002), the theoretical and legal bases for conserving large natural areas were defined in the second half of the 19th century. The landmark of this initiative was the creation of the world's first conservation unit: Yellowstone National Park, in Wyoming, USA, in 1872. This park served as a model for various Conservation Units around the world, including the Brazilian Conservation *Units*.

Over the years, other countries have joined this conservationist policy. New protected areas were created all over the world, such as *Kruger National Park* in South Africa in 1898. In 1914, Switzerland created its first park. In Canada, the first protected area was created in 1885. This was followed by New Zealand in 1894, Australia and Mexico in 1898, Argentina in 1903 and Chile in 1926 (DIAS, 2003).

In Brazil, the first protected area was the Itatiaia National Park, created in 1937. Two years later, the Iguaçu National Park was created (1939). In Rio de Janeiro, the second protected area created was the Serra dos Órgãos National Park, also in 1939, covering the municipalities of Teresópolis, Petrópolis and Guapimirim. It wasn't until 1961 that the state's third National Park was created: the Tijuca National Park (LEAL, 2004), the subject of this study.

As a way of regulating article 225, paragraph 1, items I, II, III and VII of the Federal Constitution, the National System of Nature Conservation Units was established by law number 9.985 of 18 July 2000 (SNUC). Despite generating controversy in some sectors of society, the National System of Nature Conservation Units represents a breakthrough in environmental issues, as it makes it possible to organise the law, guiding the guidelines and limitations on the creation and protection of these areas, as well as regulating the categories of conservation units at federal, state and municipal level (MITTERMEIER et al, 2005).

The SNUC has the following objectives: contribute to the conservation of varieties of biological species and genetic resources in the national territory and jurisdictional waters, protect endangered species; contribute to the preservation and restoration of the diversity of natural ecosystems; promote sustainable development based on natural resources; promote the use of nature conservation principles and practices in the development process; protect natural and little altered landscapes of outstanding scenic beauty; protect relevant geological, morphological, geomorphological, speleological, archaeological, palaeontological and cultural features; recover or restore degraded ecosystems; provide means and incentives for scientific research activities, studies and environmental monitoring; value biological diversity economically and socially; favour conditions and promote environmental education and interpretation and recreation in contact with nature; protect the natural resources necessary for the subsistence of traditional populations, respecting and valuing their knowledge and culture and promoting them socially and economically. (MINISTRY OF THE ENVIRONMENT; SECRETARIAT OF BIODIVERSITY AND FORESTS; DEPARTMENT OF PROTECTED AREAS, 2007).

In the SNUC, conservation units are defined as:

The territorial space and its environmental resources, including jurisdictional waters, with relevant natural characteristics, legally instituted by the public authorities, with conservation objectives and defined limits, under special administration regimes, to which adequate protection guarantees apply (Article 2, item I).

According to research carried out by the Ministry of the Environment, the Secretariat for Biodiversity and Forests and the Department of Protected Areas (2007), this system "consolidated a new attitude on the part of the state towards society in the field of nature conservation, creating a series of mechanisms that ensure greater public participation in the process of creating and managing protected areas" (MINISTRY OF THE ENVIRONMENT; SECRETARIAT FOR BIODIVERSITY AND FORESTS; DEPARTMENT OF PROTECTED AREAS, 2007).

The SNUC thus seeks to ensure that the creation and management of environmental protection areas are more participatory, more consistent with social dynamics and the local economy.

The Conservation Units defined and regulated in the National System of Nature Conservation Units are divided into two large groups: indirect use units, which are full protection units, and direct use units, sustainable use units (SNUC, 2000). There are five categories of Full Protection Units: Station

Ecological Reserve, Biological Reserve, National Park, Natural Monument and Wildlife Refuge. In the group of Sustainable Use Units, there are seven categories: APA - Environmental Protection Area, Arie - Area of Relevant Ecological Interest, Flona - National Forest, Resex - Extractive Reserve, Fauna Reserve, Sustainable Development Reserve and RPPN - Private Natural Heritage Reserve.

These categories can be considered typical Conservation Units, as they are part of the SNUC. There are also others - atypical - so called because, although they are covered by Brazilian law, they are not part of the Law 9.985/00 system. For example: Permanent Preservation Areas, Legal Reserves, Biosphere Reserves, Forest Servitude Areas, Ecological Reserves, listed natural monuments and Indigenous Reserves (MINISTÉRIO DO MEIO AMBIENTE; SECRETARIA DE BIODIVERSIDADE E FLORESTAS; DEPARTAMENTO DE ÁREAS PROTEGIDAS, 2007).

Atypical units are included in other rules and regulations. Even though they comply with article 225, paragraph 1, item III of the Federal Constitution, they are not included in the SNUC due to their enormous territorial dispersion, fragmentation and diversity, making it difficult to manage them in an integrated manner within the SNUC.

To finalise this introduction, let's look at some more specific issues concerning the Tijuca Forest. We will also outline some important notions regarding the development of this study. A general organisation of the work will be presented to make it easier to understand.

1.3 GENERAL CONSIDERATIONS CONCERNING THE STUDY OF PARNA-TIJUCA/RJ (TIJUCA- RJ. NATIONAL PARK).

Natural disasters threaten biodiversity, causing species extinction and habitat fragmentation. However, human action is the main cause, as it permanently jeopardises the biodiversity of the planct's ecosystems. We can therefore highlight four major threats to the survival of various existing species: The destruction, fragmentation of habitats; predatory exploitation (hunting, fishing and logging activities);the introduction of exotic species;and the increase in pests and diseases. The greatest of these dangers consists of the degradation of habitats by human actions, mainly agriculture, logging or mineral exploitation and urban occupation (PRIMACK, 1995).

Article 27 of the SNUC states that the Management Plan must cover the area of the conservation unit, its buffer zone and ecological corridors, including measures to promote its integration into the economic and social life of neighbouring communities. However, few management plans actually define the buffer zone or consider it effectively in the planning and management of natural resources.

There is no established parameter for the Buffer Zone - a term adopted by the SNUC, whose boundaries are established in the act creating the Conservation Unit or in its Management Plan. It is often confused with the concept of "surroundings", a term defined in National Environmental Council (CONAMA) Resolution 13/90, which refers to the ten-kilometre radius around conservation units

In the particular case of PARNA-Tijuca, although the buffer zone is planned and delimited, it has not yet been regulated by the government, which should have been done by decree or law. It is therefore essential that studies be carried out to identify and diagnose the external region of PARNA-Tijuca, as well as its impacts, contributing to the delimitation of the space, its proper use and the protection of the core area. We are talking about an area of great vulnerability, an urban forest that is integrated into the territory of the metropolis, making it a hostage to urban pressures.

To finalise this introduction, we will now highlight some aspects of the importance of the Tijuca Forest. Among its peculiarities, this particular forest enchants and arouses interest. It occupies a large part of the territory of a metropolis and for a long time was known as "the largest urban forest in the world".

In Rio de Janeiro, it divides the city into the south and north zones and is home to an "ecological sanctuary" along its entire length, as well as natural and historical monuments. Through the humidity it provides, it helps to soften the city's climate and to supply water through numerous water sources. Its massifs can be seen from various points in the city, including one of the city's greatest symbols, Christ the Redeemer (ABREU, 1992).

After decades of devastation, mainly caused by coffee monoculture, the Tijuca Forest is regenerating as a result of a joint effort between human action and the forces of nature. This reforestation process is a unique experience in the world, capable of structuring a secondary forest of great biological, scenic and historical value. Attracting people from all over the world, its main use by the population is therefore recreational. If we are to continue to cultivate this type of use, important conservation perspectives must not be overlooked.

CHAPTER 2

HISTORY OF THE TIJUCA FOREST

The Tijuca Forest has a rather peculiar history in terms of its formation. It has undergone a process of reforestation over almost its entire length and is now basically secondary forest. There is no other model of conservation unit in the world that has such a secondary forest profile over such a vast area.

In addition, the forest is part of a wider biome that is extremely important for biodiversity: the Atlantic Rainforest, which is considered one of the world's 25 hotspots[3] and has more than 8,000 endemic species. Despite its importance, however, its status is that of a threatened area, as less than 100,000 square kilometres (around 7%) of this native forest remain (TABARELLI, 2005).

The historical value of the Tijuca Forest is equally important. It has undergone various uses and transformations as a result of the city's dynamic history. The forest has served as an ore extraction area, a timber extraction area, a mining area, an area responsible for supplying water, an area for building quilombos, refuges for criminals and so on[4] .(HEYNEMANN, 1993).

During this specific historical period, the first reforestation efforts took place in the Tijuca massif. According to some studies in the history of science, however,

---

[3] The *Hotspot* concept was created in 1988 by British ecologist Norman Myers to solve one of the biggest dilemmas facing conservationists: what are the most important areas for preserving biodiversity on Earth? *A Hotspot* is therefore any area prioritised for conservation, i.e. one with high biodiversity and threatened to the highest degree. A *hotspot* is an area with at least 1,500 endemic plant species and which has lost more than three quarters of its original vegetation http://www.conservation.org.br/como/index.php?id=8).

[4] The writer José de Alencar was a great admirer of the Tijuca Forest. The forest inspired his great works. In 1872, José de Alencar published "Dreams of Gold", a novel set in the Tijuca Forest. The book opens by announcing the beauties of Tijuca, with the sentence: "The burning sun of February gilded the beautiful mountains of Tijuca". Also very interesting is the content of the letter the author writes to fellow writer Machado de Assis: "You know this enchanting mountain. Nature has placed it two leagues from the court, as a nest for souls tired of resting on the ground. Here everything is pure and healthy. The body bathes in crystalline waters, as does the spirit in the clarity of this blue sky. One breathes freely, not only the thin air that energises the breath of life, but that heavenly breath of the creator, who breathed into the newborn world." (ALENCAR, 1938).

there was already a cultural value of European influence, disseminated throughout society, which recognised the importance of woods, gardens and forests. Historian Heynemann (1995) emphasises that the forest corresponded to the place of aristocratic practices. One of its functions, for example, was to distract the king from his concerns of state. Most studies, however, point to the water supply crisis in the city of Rio de Janeiro as the main reason for the initiative to reforest the Tijuca Forest. Abreu (1992), in his article A cidade, a montanha e a floresta, tells us that:

There have therefore been many ways in which man and mountain have related (and still relate) in Rio de Janeiro [...] None, however, has been more important, more enduring and more problematic than the one established from the need to supply water to a city in continuous growth [...] (ABREU, 1992, p.55).

In The Garden Inside the Machine, Drummond points out:

The "mistake" in nomenclature emancipated it in 350 years from the main reason why the decimated Tijuca Forest would one day be meticulously replanted by the same society that destroyed it: the protection of the small rivers that flowed down its slopes and supplied almost all the water consumed by the city. Three and a half centuries later, the non-existence of a river of January would be more eloquent than ever (DRUMMMOND, 1988, p.280).

2.2- THE DIFFERENT USES OF THE TIJUCA FOREST OVER TIME.

Ever since Rio de Janeiro was founded in 1567 by its chief captain, Estácio de Sá, we can say that water supply problems already existed "in the city". When they expelled the French, the Portuguese conquerors chose the site between the Cara de Cão and Pão de Açúcar hills. Although it was suitable for defence, the site lacked drinking water sources. The Portuguese settlers then moved to where one of the branches of the Carioca River flowed, refuelling their boats there. This outlet became known as the "sailors' watering hole" (ABREU, 1992).

The 17th century can be characterised as a period in which we had an agricultural Rio de Janeiro. The sugar mills divided up the land in the first glebas. This was the time when the city broke the limits of the Castelo hill, expanding into the massifs (the "Centre" at the time was made up of marshes). Sugar cane plantations were the first sources of exploitation. The native trees, in turn, were used as fuel for the mills (DEAN, 1994).

With the discovery of several mines in the 18th century, Rio de Janeiro became an important emporium. Gold left the city for Portugal. In 1763, it became the capital of the colony, becoming its political and administrative centre. At the time, it already had around 40,000 inhabitants (DRUMMOND, 1997).

Due to its position in the lowlands, close to sea level, Rio de Janeiro was a city with great potential for flooding. Various diseases emerged during this period as a result of the numerous floods, which increased the unhealthiness of the city. Problems were mounting. Among them was the water supply (ABREU, 1992).

The water was supplied by fountains. The main source was the Rio Carioca, whose waters, transported by the aqueduct over the Arcos da Lapa, between the hills of Santa Teresa and Santo Antônio, supplied several fountains, four in the oldest part of the city: the Carioca, Marrecas, Moura and Largo do Paço fountains (BENCHIMOL, 1990).

In 1760, coffee was introduced to Rio de Janeiro. Coffee, or Coffea arabica, is a small tree, a rubiaceae, native to south-west Ethiopia, which has adapted well to the Atlantic Forest, spreading to various sites in the city (DEAN, 1994). In its early days, Tijuca was the place where coffee thrived the most, reaching the forest and leaving its mark.

Tijuca is a name of Tupi origin: ty-iuc, which means mud, swamp, quagmire. This term, which also means swamp, gave its name to the mountain and the entire region that arose below it. The nickname, however, originated from a lagoon on the other side of the mountain, which the Indians called Tijuca. At that time, the forest was little visited by the Indians. They feared its high altitudes because they believed it was full of spirits (DEAN, 1994).

In the province of Rio de Janeiro, coffee found a very suitable environment for its cultivation, as it thrived best in soils that were neither dry nor waterlogged. For the Atlantic Rainforest biome, however, coffee plantations posed a greater threat than any before. At the beginning of the 19th century, the place where most coffee was grown in Brazil was in Tijuca. It is from this location that it is projected into Brazilian history[5] (ABREU, 1992).

There was a belief that coffee should be planted in "virgin forest". As a result, the Atlantic Rainforest, which had been formed over thousands of years, began to be seriously threatened. Coffee is a plant that takes four years to grow and remains productive for around thirty years. All the old plantations in Rio de Janeiro were abandoned and this led to new areas of native forest being devastated in order to maintain the production of the court. Coffee advanced further and further, leaving only bare mountains (DEAN, 1994).

The big coffee growers had no knowledge of techniques to improve harvest quality and productivity. They weren't even concerned with land management. The forest was cleared almost entirely in preparation for planting. Some of the felled trees were used to make charcoal and supply the city, while others were used as building material (DEAN, 1994).

With these practices, farmers obtained a low yield per hectare planted. They didn't worry about conserving the soil's natural resources, causing the coffee trees to decay after about twenty years of growth. If coffee planting had been done more carefully, a large part of the Atlantic Rainforest would possibly have been preserved (DEAN, 1994).

The plantations spread to various farms in the city. When the Royal Family arrived in 1808, coffee already occupied a large part of the Tijuca massif, especially in the Alto da Boa Vista area, where it adapted very well (DEAN, 1994).

The royal family brought with them no less than 20,000 people, including nobles, soldiers, officials, priests and their families. It was a dizzying growth in population, around 25%. The famous "transmigration of the royal family" meant that Rio de Janeiro experienced the ambiguous situation of being both the colonial

---

[5]In the 19th century, as is well known, coffee became the country's main economic activity. According to some authors, the Brazilian state was saved by the coffee revenues that converged on the customs house in Rio de Janeiro. It saved both the colony's aristocracy and the imperial court. The product of the large estates became coffee. However, at the same time as it was largely responsible for the country's growth, coffee plantations devastated the forest. It is also important to point out that, in its early days, slave labour generated the strong economy that coffee represented (ABREU, 1992).

capital and the seat of the Portuguese Empire (ABREU, 1992).

Nobles, military personnel and officials (as well as clergymen) were given the privilege of occupying the best buildings and residences in the still small city. This presence was a great shock to Rio's social life. The Tijuca mountain area, already penetrated by the estates of members of the local and foreign elite, was naturally an attractive option for the new, powerful and wealthy inhabitants.

A curious example relates to the arrival of the French Artistic Mission in 1816. Taunay, Debret and Grandjean de Montigny's mission was to bring the neoclassical style to the Court in the tropics (HEYNEMANN, 1993). The passage of this group through the city brought together new owners and artists in the Tijuca Forest. According to Maya (1967), it was only after the painter Nicolas Antoine Taunay bought a site opposite Cascatinha da Tijuca and settled there with his family that the area began to be mentioned as a place of great natural beauty and a favourable climate.

The presence of the elite in the locality was definitely attested to by the nickname the area earned in the mid-1820s: Tijuca Imperial. Its farms, the oldest and the most recent, occupied most of the slopes of the Tijuca mountain range (DRUMMOND, 1988).

However, not everything was going marvellously in this scenario. One of the serious problems of the period was how to supply water to so many people if the only existing water source was insufficient. The intensification of the supply crisis lasted throughout the Johannine period[6] and continued throughout the Imperial period (ABREU, 1992).

In addition, in 1814, some quilombolas were captured in Tijuca. Nobody knows for sure when these quilombos were built, only that they already existed when the court arrived. The Tijuca Forest, being a mountainous area covered in forest, with many caves and streams, was home to the largest and most important quilombos in Rio de Janeiro.

In an attempt to solve some of these problems, King João VI issued a decree

---

[6]After the arrival of the Portuguese royal family in Brazil, the Johannine period began (18081821).

in 1817 ordering an end to the felling of trees near water sources and on the banks of streams close to the capital (DRUMMOND, 1997). Even so, until the 1840s, coffee plantations increasingly dominated the Tijuca Massif (ABREU, 1992).

Coffee planting was therefore associated with a reduction in the availability of water, which even changed the generalised rainfall regime in Rio de Janeiro. The city was hit by severe droughts in 1824, 1829, 1833 and 1843. The latter, which was very severe, caused the population to resort to ships anchored in Guanabara Bay to obtain water (ABREU, 1992).

The drought of 1824 revealed that the reinforcement achieved by piping water from the Maracanã River was still far from solving the city's water problems. With the drought of 1829, the Intendência de Polícia began searching for new water sources. In 1833, six more springs were found in the Paineiras region, which would have their water led to the Carioca pipeline. The destruction of the vegetation around the Serra da Carioca springs, combined with less abundant rainfall, produced major deficits in the supply of drinking water (DRUMMOND, 1997).

In 1843, the coffee crop began to decline due to a plague that caused productivity to fall. In addition, the lack of proper soil management and care brought the coffee crisis to an early stage. In Abreu's words:

Planted without any concern for maintaining the fertility of the soil, coffee plantations soon accelerated the process of depleting the land. What's more, as they climbed the slopes vertically (instead of following the contour lines), coffee plantations also began to contribute decisively to the acceleration of erosion processes, as they made it easier for rainwater to wash away the topsoil. (ABREU, 1992, p. 76)

In 1843, the Imperial Government formed a commission to suggest measures to protect water sources. It was at this time that the conservation of the forests of the Tijuca massif was proposed. In 1845, the same commission announced:

[...] to penalise people who cut down and burn at the headwaters and slopes of the Carioca and Maracanã rivers; at the tops of the mountains and their slopes, and along the respective pipelines" (MINISTÉRIO DO IMPÉRIO/1844).

As the population grew and economic activities expanded, so did the general demand for water. Few measures, however, had actually been taken to preserve the riparian forests of the springs and rivers that supplied the city. In 1850, the new Maracanã River pipeline came into operation, expanding the city's water supply.

There was a reforestation plan at the time, but it wasn't fully implemented. It was a very slow process. The proposal for expropriations didn't come to fruition until 1860, when Rio de Janeiro's population was already around 400,000. At this time, another drought hit the city. Scholars understood that preserving the springs located in the Tijuca massif was a fundamental condition for solving many of the city's problems, especially water supply (ABREU, 1992).

Contrary to popular belief, the occupation of the hillsides began around 1859 by people of high social status. The reason was the unhealthy conditions in which the centre of Rio de Janeiro found itself. The city was cut off by ditches that were constantly clogged with waste thrown up by the population. Everything was thrown into the streets, which were still unpaved.

The situation worsened with the yellow fever epidemic and other contagious diseases such as cholera, which claimed a large number of lives. This strengthened the trend - which had already begun - of occupying the areas at the foot of the Serra Carioca, such as Botafogo and Tijuca. These localities then saw their old estates cut up even further, especially with the opening of new streets and the demarcation of urban plots. This process of urban occupation of the mountains of Rio de Janeiro, which began in the mid-19th century, continues to this day (ABREU, 1992).

Abandoning the "stinking city" was a common goal among those who could enjoy the privilege of being away from the central areas of Rio de Janeiro. These epidemics of yellow fever and cholera would have a marked effect on the city. The Tijuca region was considered a good refuge from these calamities.

In 1853, Luís Pedreira do Couto Ferraz, the Viscount of Bom Retiro, a resident of Tijuca, played a significant role in the defence of the region. It was during his tenure in the Ministry of King Pedro II that appraisals were carried out on private land to be acquired. The expropriations only came in 1855.

The imperial government acquired a small number of properties strategically located near the springs and upper reaches of the Carioca, Maracanã and Comprido rivers. In the meantime, the issue of water supply was transferred to the newly created Ministry of Agriculture, Commerce and Public Works (DRUMOND, 1997).

Between 1855 and 1856, a devastating cholera epidemic hit Rio de Janeiro. During this period, the problem of forest destruction in Rio de Janeiro was already being addressed by members of various intellectual circles. Planting trees was considered a necessity, an urgency for Rio de Janeiro. The process of expropriating land to avoid deforestation began, prioritising areas where there were water sources (ABREU, 1992).

Even with the investment in water resources, such as the new pipes in Maracanã, Andaraí Grande and Cabeça, water shortages were still common in the city. In 1860, the Imperial Government proposed a *General Water Supply* Plan, due to the significant decrease in the springs that supplied the city. The government appointed a specific commission to draw it up (ABREU, 1992).

A great controversy then began. Some believed that the water sources close to the city were no longer capable of satisfactorily supplying the city. They argued in favour of immediately supplying and channelling the rivers that flowed down from the Serra do Tinguá.

Other engineers, on the other hand, defended the idea that the rivers that flowed down from the Tijuca massif could still meet the water supply needs of the city of Rio de Janeiro. Therefore, for this to happen, all that was needed was for the water sources to be preserved.

The commission's final solution ended up strengthening the ideas of those who favoured maintaining the water supply through the Tijuca massif. The commission understood that the preservation of water sources was a fundamental condition for solving many of Rio de Janeiro's problems, including flooding.

In 1861, Emperor Pedro II appointed Major Manuel Gomes Archer as *Administrator of the Tijuca Forest.* The process of forest restoration began, with an initial team of six slaves and then more than twenty-two salaried workers. Archer lived on the Independência farm in Guaratiba. His title of major is of unknown provenance. According to his own account, at the time he only had practical knowledge of forestry (HEYNEMANN, 1993).

In the cultivation of these trees, the major made use of sowing and the formation of nurseries, announced as the only method that would fulfil the government's objective: "to fill the clearings of the forest with leafy trees within a few years, the future deposit of wood for civil and naval construction" (HEYNEMANN, 1993). Other results besides obtaining timber were also included, such as improved health conditions and the influence of forests on climate and rainfall (HEYNEMANN, 1993).

Also in 1861, the waters of the Taylor River were abstracted in Alto da Boa Vista. In 1863, the Public Works Inspectorate advised on the urgent need to utilise the waters of the Trapicheiro River by piping the Maracanã River. In 1867, the Imperial Government acquired more land, expanding the potential area of the reforestation project.

In 1868, a new drought hit Rio de Janeiro, making it urgent to draw on some springs that had not yet been used in the upper Tijuca (Cachoeira river), Andaraí Grande (Joana) and Jardim Botânico (Macacos river and tributaries), which became part of the city's water supply system.

In 1870, a new commission was created by the Imperial Government. This commission was headed by engineer André Rebouças. The commission's suggestion once again raised the possibility of collecting water from the Serra do Tinguá as an alternative to the problem of water supply. Due to further droughts, five years later the Imperial Government finally decided to adopt this measure.

The works to collect and distribute water from the Ouro and Santo Antônio rivers in the Serra do Tinguá began on 12 September 1876. They were provisionally inaugurated in the presence of Emperor Pedro II in May 1880. Also in the 1870s, the waters of the tributaries of the Gávea river (near Pedra Bonita), the small water springs that came down from the Serrado Andaraí Pequeno (from the Férreas waters to Trapicheiro) and the Covanca would be channelled.[7]

Another consequence of this new water catchment option was a certain lack of interest in the Tijuca Forest on the part of the public authorities. Far from Rio de Janeiro, Tinguá's water sources were also more abundant than those of the Carioca, Maracanã and Comprido rivers. As a result, the importance of these three rivers for the water supply of the city of Rio de Janeiro diminished more and more.

In 1873, Archer signed the report of the Tijuca Forest Service. In his final balance sheet, he established the number of 61,852 saplings planted during his tenure. There are assumptions, however, that he planted 20% more and that the figure of 72,000 seedlings planted is a reasonable estimate. In this same report, Archer demonstrates the importance and urgency of legislation and the creation of forestry institutes for scientific learning, which is made clear in the passage: "... greater and better commitment to methodical forestry work, not only here in the vicinity of the Court, where it is undeniably and urgently needed, but also throughout the rest of the Empire..." (Report by the Administrator of the Tijuca National Forest, attached to the report by the Ministry of Agriculture, Commerce and Public Works. Rio de Janeiro. 1874)

Archer evidently conceived the replanting of the Tijuca Forest with the aim of providing permanent protection for Rio de Janeiro's water sources and not as a temporary crop of trees to be cut down later (according to the German "forestry engineering" conception, of which Archer himself seems to have been a scholar) (DRUMMOND, 1997).

In 1874, Archer resigned from his post, citing the need for more workers. In 1877, Gastão de Escragnolle took over as Administrator of the Tijuca Forest. He continued the task of replanting the trees, including exotic plants (from other continents), although on a smaller scale and at a slower pace than Archer.

This administrator dedicated his administration to the task of making the Tijuca Forest accessible to visitors. He wanted to turn the forest into a place of

---

[7] Rio de Janeiro needed this new source, as its population had almost doubled since 1872, from 275,000 to 522,000 inhabitants in 1890 (DRUMMOND, 1997).

leisure and recreation. He opened roads, parks, belvederes, fountains, trails, bridges and artificial lakes in the forest. Escragnolle also had the help of the French naturalist and landscaper Auguste Glaziou in his work (IBASE, 2005).

Restoring water sources remained an important task. During Escragnolle's administration, between 1877 and 1887, 21,500 new seedlings were planted. It is estimated that the Tijuca Forest was re-established between 1862 and 1887 with around 100,000 seedlings.

2.3- THE ARRIVAL OF THE 20TH CENTURY.

The end of the 19th century saw the emergence of tourism as an important economic activity for the city's development, which would become even stronger in the 20th century. The valorisation of the mountain as a healthy place was of great importance in this new perspective. The increase in demand for the Alto da Boa Vista region and a series of investments made in the Tijuca massif prove this trend, especially with the construction of the Corcovado Railway and the Paineiras Hotel, awarded to engineers Francisco Pereira Passos and João Teixeira Soares. Both developments were inaugurated on 9 October 1884 in the presence of Emperor Pedro II and the entire Imperial Family (ABREU, 1992).

It is important to mention, among the remarkable facts of the time, that on 15 November 1889 the republic was proclaimed in Brazil. Unfortunately, as is common in our political tradition, especially at times of transition, the characteristic actions of the old governments are set aside. Thus, many acts of the Imperial Government's management came to be considered unimportant by the creators of the new regime. As a result, the Tijuca Forest experienced almost half a century of considerable administrative neglect (1889 to 1943).

At the beginning of the 20th century, due to the urban reform of the city centre led by Mayor Francisco Pereira Passos, the way the city was occupied changed. The lower classes began to occupy some of the city's hills. The first favela formations in Rio de Janeiro were Morro da Providência in 1897 and Morro Santo

Antônio in 1893, both in the city centre.

The government demanded housing regulations, but this was unfeasible for the lower classes. Since then, no concrete alternative housing policy has been proposed, which has only increased the occupation of Rio's hillsides. By 1920, the favelas had spread to Botafogo (Pasmado), Copacabana (Tabajaras and Leme), São Cristóvão (Mangueira).

During this period, irregular occupation also began in the Tijuca Massif: on the slopes of Serra Carioca (São Carlos and Querosene), Tijuca (Salgueiro) and Engenho Novo (Macaco). Occupation was also intense on the southern slopes (Rocinha and Dona Marta).

From 1930 onwards, with accelerated growth and industrial expansion, the slopes of the Tijuca massif became densely populated. These slopes became intensely occupied by the middle and upper classes in the Jardim Botânico, Gávea and Cosme Velho regions, who were looking for milder climates. Access at this time was already by car (ABREU, 1992).

The growing occupation of the slopes of the Tijuca massif (by the rich, the middle class and the poor) has put an old problem on the city's agenda: flooding. As the population grew, the

The intensity of occupations also increased. The opening of roads such as Avenida Edson Passos (which gives access to Alto da Boa Vista), the Lagoa-Barra motorway and the Grajaú-Jacarepaguá motorway further intensified urbanisation (ABREU, 1992).

At the end of the 1920s, the city council hired Alfred Agache to draw up the city's Master Plan. One of the suggestions in the Master Plan was the creation of the Rio de Janeiro National Park, intended to be a constant reservoir of air, water and vegetation.

In 1943, the mayor of Rio de Janeiro, Henrique Dodsworth, summoned Raymundo Ottoni de Castro Maya to be the new administrator of the Tijuca Forest and to restore it. He restored most of the gardening and landscaping work carried out by Gastão de Escragnolle and Auguste Glaziou, which had been covered up by the forest and the contempt of the public authorities. Parks, squares, trails, houses, lakes and other facilities were in terrible condition.

Castro Maya felt rewarded for having significantly increased the number of visitors attracted to the Tijuca Forest during his tenure (1943-1947). He estimated that on Sundays there were around five thousand visitors to the Tijuca Forest. The Mayrink Chapel, for example, dating from 1860, was remodelled in 1943 during his administration. Its altar contains paintings by Cândido Portinari, donated by residents of Alto da Tijuca.

In 1960, the Tijuca Forest came under the administration of the newly created state of Guanabara. On 6 July 1961, the federal government created the Tijuca National Park, including the Tijuca Forest, by Decree No. 50,923. In fact, the official name of this Conservation Unit was, at the time, the Rio de Janeiro National Park. Between 1959 and 1961, the federal government created eleven national parks. In this way, Raymundo Ottoni Castro Maya's project to turn the Tijuca Forest into a National Park was realised (HEYNEMANN, 1993).

In 1966, the Tijuca National Park (still called the Rio de Janeiro National Park) was listed by the National Historical and Artistic Heritage Institute (IPHAN). On 28 February 1967, Decree No. 60.183 redefined the spatial limits of the park, changing its official name to Tijuca National Park. Part of the forest in the Covanca and Andaraí areas was excluded from the park because several favelas were located on its slopes.

In 1981, the Tijuca National Park Management Plan was published. A fundamental point addressed in this plan was zoning, since the park has been under enormous pressure from the metropolis that surrounds it. The latest Management Plan[8] was published in 2008.

It is currently intended to carry out a partial review of this plan, reformulating the activities planned and carried out, as well as updating the unit's planning. However,

---

[8] A management plan is a mandatory document that every Conservation Unit must draw up. It is defined in article 2, item XVII, of the SNUC/Law 9,985/2000, as: XVII - management plan: a technical document which, based on the general objectives of a conservation unit, establishes its zoning and the rules that should govern the use of the area and the management of natural resources, including the implementation of the physical structures necessary for the management of the unit.

no deadline has yet been set for such a review[9] .

CHAPTER 3

THE TIJUCA FOREST TODAY

The Tijuca Massif occupies a prominent place in Rio de Janeiro, as it is the great divider of the city. In fact, its growth and expansion were dictated to a certain extent by the presence of the mountains and the proximity of the sea (ABREU, 1992).

The relationship between nature and Rio society is historically constructed. The basis of this construction, in the case of the Tijuca Massif, is related to the growing occupation of its slopes by the population and the use of natural resources, described in the previous chapter, which continues to this day.

The municipality of Rio de Janeiro is home to three large forest remnants: the Tijuca Massif, Pedra Branca and Serra do Mendanha-Gericinó. These remnants are included in the Atlantic Forest Biosphere Reserve (LINO, 1992; SEMA, 2001) and have marked biological similarities (faunal and floral), as a result of their considerable geographical proximity and the fact that they were continuous areas in the past (ROCHA et al, 2003).

The climate is tropical, with an average annual temperature of 22° C and annual rainfall of around 2,300 mm (COELHO NETTO, 1992). The predominant vegetation is secondary Ombrophilous Forest[10] , without palm trees and in an advanced state of regeneration (IBGE, 1993). These forests are currently being protected by legal instruments that have transformed them into conservation units for integral protection and sustainable use at the municipal, state and federal levels.[11]

10Dense ombrophilous forest is an evergreen forest with a canopy of up to 50 metres, with emergent trees up to 40 metres high. It has dense shrub vegetation, made up of ferns, arborescents, bromeliads and palm trees. Climbers and epiphytes (bromeliads and orchids), cacti and ferns are also very abundant. In humid areas,

The term created by Ellemberg & Mueller-Dombois replaced Pluvial (of Latin origin) with Ombrophilous (of Greek origin), both with the same meaning "friend of the rains". Its main ecological characteristic lies in the ombrophilous environments, related to the higher thermo-pluviometric indices of the coastal region and the Amazon. The well-distributed rainfall throughout the year determines a bio-ecological situation with practically no dry period (0 to 60 days a year).

The areas of the Tijuca Massif make up PARNA Tijuca, created in 1961. Adjacent to the Park is the Grajaú Forest Reserve, transformed into the Grajaú State Park (State Decree NQ 32.017/2002). This partly increases the protective belt of forest contained in the National Park (ROCHA et al, 2003). To the west of the Tijuca National Park and covering the region of the Pedra Branca massif, lies the largest remnant of Atlantic Forest in the municipality, with an area approximately four times larger than the Tijuca National Park. It is now the Pedra Branca State Park, which includes the parcels of land above 100 metres above sea level up to the 1,024m Pedra Branca peak, the highest point in the region.

Within the Tijuca Forest there is an important hydrographic network (eight basins and 53 micro-basins). This partly supplies the surrounding resident population. Similar to the Tijuca Massif region, the Pedra Branca Massif area, as a result of the continuous process of degradation of forested areas caused by the former presence of several colonial coffee plantations, was identified at the beginning of the 20th century as relevant for the preservation of Rio de Janeiro's water resources, which became a determining factor in government priorities for the protection of water sources (ROCHA et al, 2003).

The growing occupation of farms growing various crops (oranges, coffee and bananas) in the massif came to represent a continuous risk for the entire region, culminating in the emergence of social movements for the preservation of the area and the institution by the federal government of the Union Protection Forests at the beginning of the 20th century (Rocha et alii, 2003).

Most of the Tijuca Massif is home to the Tijuca National Park, which is located in the centre of the city of Rio de Janeiro, in the mountains of the Tijuca Massif, between parallels 22°55'S and 23°00'S and meridians 43°11'W and 43°19'W, in the south centre of the state of Rio de Janeiro. The massif is characterised by its rugged terrain, oriented in a NE/SW direction, and also includes the Serra do Mar fault block,

---

ELLEMBERG, H.: MUELLER- DOMBOIS, D. A. Tentative physiognomic-ecological classification of plant formations of the earth. Separate part of Ber. Geobot. Inst. ETH, Zurich. 1965/66. *apud* IBGE - Instituto Brasileiro de Geografia e Estatística.
[11]As we saw in the introduction and in chapter 1 of this work.

whose altitude varies between 80m and 1,021m.

Geologically, it is mostly made up of gneisses, with some granitic intrusions (Helmbold et al, 1965).The relief and soils are related to the tectonic faults of the Tertiary, predominantly latosols with the occurrence of lithic neosols and cambisols, which form a weathered mantle, shallow in the steeper stretches and several metres deep in the valley bottoms (Coelho Netto, 1992).

The presence of facoidal gneiss is responsible for peculiar morphological features such as pontoons (popularly known as "sugar loaf"), steep walls, peaks, mesas and a whole morphological ensemble that has come to symbolise the city of Rio de Janeiro (COELHO NETTO, 1992).

On 8 February 1967, Federal Decree 60.183 changed the name to Tijuca National Park and defined its boundaries, with three areas separated by public roads: Tijuca Forest (Sector A), Corcovado-Sumaré- Gávea Pequena Set (Sector B) and Pedra Bonita-Pedra da Gávea Set (Sector C). By means of Federal Decree s/n of 3 July 2004, PARNA-Tijuca's boundaries were amended and extended to include Parque Lage and Sector D, made up of Covanca /Pretos Forros.

After the creation of the current PARNA-Tijuca and with the increase in urban pressure, Decree 322/1976 established new rules for housing construction. Thus, a protection strip was created around the park, encompassing its massif, and construction was prohibited at points above 100 metres in height, known as cota 100.

This decree was followed by other decrees regulating various areas already occupied by the massif. One of them, on the Furnas - Edson Passos axis, recommended the creation of the Alto da Boa Vista environmental protection and urban recovery area (APARU), part of whose area overlaps to a large extent with the limits of PARNA-Tijuca.[12]

Sector A, the Tijuca Forest, covers an area of 14,732,718.68 square metres, or 1,473.27 hectares. It is made up of the Andaraí, Tijuca and Três Rios Forests and is the area most visited by the local population. It has a main access called the Cascatinha

---

12 This APARU issue will be further detailed in the next chapter.

Gate and an exit gate located next to the Solitude Dam. The Tijuca Forest includes several areas of intensive use, including restaurants, waterfalls, dams, leisure areas, trails, caves and viewpoints. There are also historic buildings in this sector, such as the Mayrink Chapel, the Barracão (where the PNT's administrative headquarters are located), the old Hípica campestre and several 19th century ruins. This sector is close to the Usina, Andaraí, Grajaú, Jacarepaguá, Alto da Boa Vista and Itanhangá neighbourhoods. Enforcement in these areas is sporadic and almost always depends on reports of offences (PLANO DE MANEJO/ PARNA- Tijuca, 2008).

Sector B, Serra da Carioca, has an area of 117,100.82 square metres and a perimeter of 1,764.07 metres. This sector is home to the Christ the Redeemer monument. Travel in this sector is via a series of roads, such as: Redentor, Vista Chinesa and Sumaré. Located at the top of the mountain range of the same name, it separates the south from the north of the municipality of Rio de Janeiro. It neighbours the neighbourhoods of Usina, Muda, Tijuca, Rio Comprido, Santa Teresa, Cosme Velho, Botafogo, Humaitá, Jardim Botânico, Gávea, São Conrado and Alto da Boa Vista.

This sector is crossed by the Sumaré, Redentor, Corcovado, Paineiras, Vista Chinesa and Dona Castorina roads. There were five vehicle access gates in this area: Passo de Pedras, Macacos, Caboclas, Sumaré and Sapucaias. However, due to a lack of workers to man the gates, they were deactivated and demolished, leaving only one gate and the guardhouse for collecting entrance tickets to Corcovado, under the responsibility of an outsourced company. This circuit that gives access to Christ the Redeemer is made by the Corcovado railway.[13] In addition to the Corcovado Train, a line of vans has been in operation since July 2013, linking Largo do Machado to Corcovado. This transport is carried out by the Paineras Corcovado company, which

---

[13]The Corcovado Railway was the first electrified railway in Brazil. Inaugurated in 1884 by King Pedro II, it is older than the Christ the Redeemer monument itself. In fact, it was the railway that transported the pieces of the monument for four consecutive years. At the time, the steam train was considered a miracle of engineering, as it covered 3,824 metres of railway line over totally steep terrain. But in 1910, the trains were replaced by electric machines and more recently, in 1979, when the Esfeco company took over the railway, more modern and safer models were brought from Switzerland (Teixeira et al, 2012).

won the tender promoted by ICMBio (Instituto Chico Mendes para a conservação da Biodiversidade/ ICMBio, 2013).

Christ the Redeemer, inaugurated on 12 October 1931, is one of the main symbols of the city of Rio de Janeiro. Recently, in a vote organised by the New Seven *Wonders Foundation* in Switzerland, via the internet and mobile phone messages, it was voted one of the Seven New Wonders of the Modern World, out of twenty-one participating monuments from all over the planet (Instituto Chico Mendes para a conservação da Biodiversidade/ ICMBio, 2013).

In addition to Alto do Corcovado, this area has other important tourist attractions in the city of Rio de Janeiro, such as: Mirante Dona Marta, Mirante Andaime Pequeno, Mirante Bela Vista, Mirante Barro Branco, Mirante Vista Chinesa, Mesa do Imperador and Mirante Curva dos Bonecos. The Serra da Carioca is monitored through weekly patrols and in response to reports of irregularities (PLANO DE MANEJO/ PARNA-Tijuca, 2008).

One of the most representative works of the past, an integral part of the historical heritage of the city of Rio de Janeiro, is Parque Lage[14] , which has also been part of the Serra da Carioca sector since 2004. Its buildings are currently used by the School of Visual Arts (EVA) and the NGO Renascer. In terms of surveillance in this area, the Municipal Guard provides support during the day, while an outsourced company controls the entrance gate for 24 hours. (PLANO DE MANEJO/ PARNA- Tijuca,

---

[14]With regard to Parque Lage, it is known that it was originally part of the Engenho de Nossa Senhora da Conceição da Lagoa, which also gave rise to the Botanical Garden. The park area was acquired in 1611 by the family of Rodrigo de Freitas Mello, who made the area his official residence. It was only in 1859 that the property was sold to Comendador Antônio Martins Lage and became known as "Chácara Lage". The Lage family carried out various improvements and transformed the residence into an eclectic-style mansion, designed by the Italian architect Mário Vodret. In 1941, the farm was sold to a property company that decided to knock it down to build a complex of buildings. As a result, in 1957 demands began to be made for the preservation of its historical and cultural heritage. It wasn't until 1965 that the area was listed by the then State of Guanabara and, finally, in 1976, the Chácara Lage lands were expropriated and became the property of the Federal Government.(*Tijuca National* Park Management Plan - *Part 3: Analysing the Conservation Unit).An* important step in relation to the Park's land situation was taken with the publication of Federal Decree s/N° of 4 July 2004, in which the last article states that the entire area of the Tijuca National Park should be transferred from the Union Patrimony Service (SPU) to IBAMA, now the Chico Mendes Institute for Biodiversity Conservation (ICMBio), with a view to taking possession of the land under its responsibility.

2008).

Sector C, the Pedra da Gávea - Pedra Bonita complex, is part of the Tijuca Massif. Its slopes face the neighbourhoods of Gávea Pequena, São Conrado, Barra da Tijuca and Alto da Boa Vista. Access to Pedra Bonita is via Estrada das Canoas and to Pedra da Gávea via Rua Sorimã. Supervision in this sector is poor and in the Pedra da Gávea area, where there is a high frequency of visitors, there have been reports of offences. It has an area of 2,578,421.16 square metres and a perimeter of 10,117.12 metres. Of all the sectors of the PNT, due to its geomorphological (rocky) characteristics, it is the one with the most impressive features. This sector is very popular with sports enthusiasts, such as free flight, mountaineering and hiking.

Sector D, the Serra dos Pretos Forros and Covanca complex, is located to the north of the Tijuca Massif and is bounded by the Menezes Côrtes road, but also by the north of the city and Jacarepaguá. Access is via the upper part of the aforementioned road and Estrada da Covanca. Its area is 4,782,299 square metres, or 478.22 hectares.

It currently has vegetation cover in various stages of regeneration and some areas of natural vegetation. The existence of this area is very important for maintaining the ecosystem and increasing the viability of fauna and flora populations. Expropriated at the time of the Empire with the aim of protecting the water sources of the city of Rio de Janeiro, Covanca and Pretos Forros were part of the Rio de Janeiro National Park under Decree 50.923 of 6 June 1961. However, on 8 February 1967, when Federal Decree 60.183 changed its name to Tijuca National Park and established the current boundaries, this area was not included in PARNA-Tijuca. The region, located to the north and north-west of the Tijuca Forest sector and separated from it only by the Grajaú-Jacarepaguá road, is an extension of PARNA-Tijuca.

According to the Management Plan, Covanca has been defined as an environmental recovery area, since there is no intention at the moment to intensify visitation in this area - above the Yellow Line, Jacarepaguá - which is already very degraded and under pressure from favelisation. This sector was incorporated into PARNA-TIJUCA in 2004. Inspection of the area is still problematic, as it is surrounded by communities where criminal gangs are present.

Today, hundreds of favelas border the Tijuca Massif. The following favelas are noteworthy because of their proximity to the UC boundary: in the Serra da Carioca-Rocinha (São Conrado/Gávea), Dona Marta (Botafogo), Guararapes (Cosme Velho), Coréia and Formiga (Tijuca) sectors; in the Floresta Sector - Borda do Mato/Nova Divinéia (Andaraí/Grajaú); and in the Pretos Forros/Covanca sector - the Inácio Dias favela (Jacarepaguá).

CHAPTER 4

RESEARCH *METHODS*:

The aim of this study was to assess the relationship between some of the communities located in the buffer zone, surroundings and adjacencies of the Tijuca National Park (PARNA-Tijuca), to identify the pressures resulting from vacancies in these areas and from visitation, and to survey the main environmental impacts.

The methodology used in this research consisted of direct observation at some points in the surroundings, buffer zone and adjacencies of PARNA-Tijuca. Field observations were aimed at identifying the most significant environmental impacts and bibliographical research. Structured questionnaires were used to collect historical data and local research. This also helped to point out other issues, in addition to the most significant environmental impacts, which are highly relevant.

The questionnaire was also adapted to be applied in the four sectors of PARNA Tijuca (Sector A - Tijuca Forest; Sector B - Carioca Mountains; Sector C - Pedra da Gávea/Pedra Bonita; Sector D; Pretos Forros/Covanca) both in the core area and in the external areas. And in the final part, the questionnaire is aimed exclusively at residents of the buffer zone, surrounding areas and neighbourhoods.

Most of the data in this study comes from interviews, via a standardised questionnaire, and relevant additional questions with the various actors involved. Most of the questionnaire used closed questions, with programmed and quantifiable answers, and other questions were asked according to some more specific characteristic of the planned profiles.

The sample totalled ninety people with very heterogeneous profiles who were related to PARNA-Tijuca at some level.

Throughout the research, it was observed that outside the boundaries of the Tijuca National Park Buffer Zone (PARNA-Tijuca), there are dense urban settlements that have significant environmental impacts on an area of great natural potential.

Therefore, there is a great diversity of fauna, flora, speleological riches[15] , historical monuments and water sources. As such, the buffer zone proposed by the PARNA-Tijuca Management Plan is very restricted, as it is located very close to the Conservation Unit boundary. For this reason, the research extends its studies to part of these areas. Referring to them as adjacencies.

The PARNA-Tijuca Management Plan[16] uses the concept of the surrounding area as its definition of the buffer zone and uses CONAMA resolution 13/1990 as a reference, which classifies the surrounding area as a 10 km strip from the boundary of the Conservation Unit. However, according to the SNUC, the delimitation of the Buffer Zone must be established in the Conservation Unit's Management Plan, which even though it also uses the term "surrounding area" makes no reference to the definition in CONAMA resolution 13/90. As such, the SNUC does not explicitly stipulate that the Buffer Zone must be delimited; the term "surroundings" was merely adopted. This is clear from two articles, article 2, item XVIII, which defines the Buffer Zone: "the surroundings of a conservation unit, where human activities are subject to specific rules and restrictions, with the purpose of minimising negative impacts on the unit." (BRASIL, 2000). And also in article 25, which states:

Conservation units, with the exception of Environmental Protection Areas and Private Natural Heritage Reserves, must have a buffer zone and, where appropriate, ecological corridors.

§ The body responsible for administering the unit shall establish specific rules regulating the occupation and use of resources in the buffer zone and ecological corridors of a conservation unit.

§ Paragraph 2 The boundaries of the buffer zone and ecological corridors and the respective rules referred to in Paragraph 1 may be defined in the act creating the unit

---

[15] A group of natural cavities, such as caves and caverns.

16 Management plan: a technical document which, based on the general objectives of a conservation unit, establishes its zoning and the rules that should govern the use of the area and the management of natural resources, including the implementation of the physical structures needed to manage the unit. Source: SNUC/2000 article 2, item XVIII.

or later. (2000).

An analysis of the two articles of the SNUC mentioned above leads us to conclude that what makes it possible to expand or restrict the boundaries of the Buffer Zone are criteria that will be in accordance with the needs of preservation, the objectives of use and the positions of the management, advisory and deliberative bodies of the CU.

The concept of environmental impact will be guided by CONAMA Resolution No. 001 of 23 January 1986 in its Article 1, which contains the following definition:

[...] any alteration of the physical, chemical and biological properties of the environment, caused by any form of matter or energy resulting from human activities that directly or indirectly affect: I- the health, safety and well-being of the population; II- social and economic activities; III the biota; IV- the aesthetic and sanitary conditions of the environment; V- the quality of environmental resources.

The questionnaires were administered in the following ways: direct interviews, in the core area, buffer zone and adjacent areas of sectors A, B and C of PARNA-Tijuca, via the internet, and printed questionnaires delivered to some people, who answered them at home and then handed them in. The differences reflected in the way the questionnaires were administered and the responses obtained were subtle. Some interviewees who answered via the internet, even though they lived far away from the

PARNA-Tijuca, residents of the ZA were considered. Most of the interviews were carried out in the core area of sector A, which is where the visitor centre and most of the staff are located. There was no field research in sector D, which is why there were no interviews there. Although people who live in the neighbourhood of this sector are part of the sample. Until now, sector D has been earmarked for environmental recovery and there is no structure for visitors. But the main reason for not carrying out field research in the surrounding area is due to the high level of violence generated by drug trafficking in the communities located there (Figure 1).

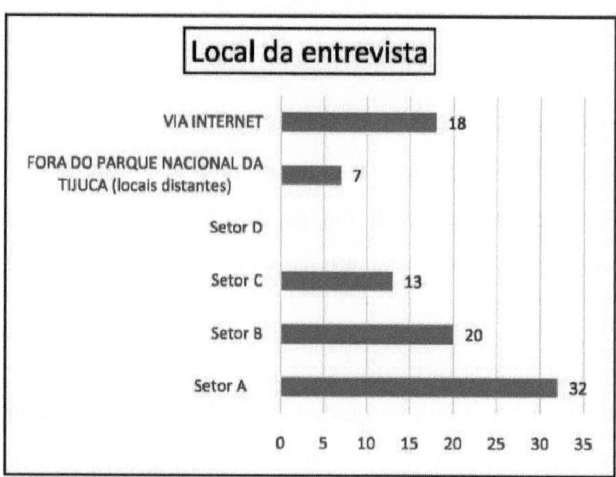

Figure 1 - interview location.

CHAPTER 5

RESULTS AND DISCUSSION.

Most of the people who took part in the sample are residents of Rio de Janeiro and most of the interviewees who live outside the state of Rio de Janeiro or in another country were interviewed in Cristo Redentor (sector B, core area). The age range also varied, from twelve to eighty-one. The predominant age group was between forty-one and sixty, which accounted for thirty-four of the sample (38 per cent). It's important to emphasise this in order to understand whether the answers varied between the generations, both in terms of environmental information and their view of nature. There was no defined profile of responses according to age group, which would show, for example, that people in higher age groups had more information. As such, it can be seen that age is not related to knowledge about PARNA-Tijuca, but rather to access to education, in the case of occasional visitors, and the involvement that develops when living in the area, in the case of residents, employees, volunteers and more regular visitors.

In terms of level of education, the sample covers all levels of education. From no schooling to a doctorate. However, there is a predominance of people with completed higher education, which corresponds to 25% of the ten education categories included in the questionnaire. This data corresponds to the survey sample, which includes visitors and employees of PARNA-Tijuca, as well as local residents. The data would possibly be different if only the residents of the surrounding communities had been part of the sample (Figure 2).

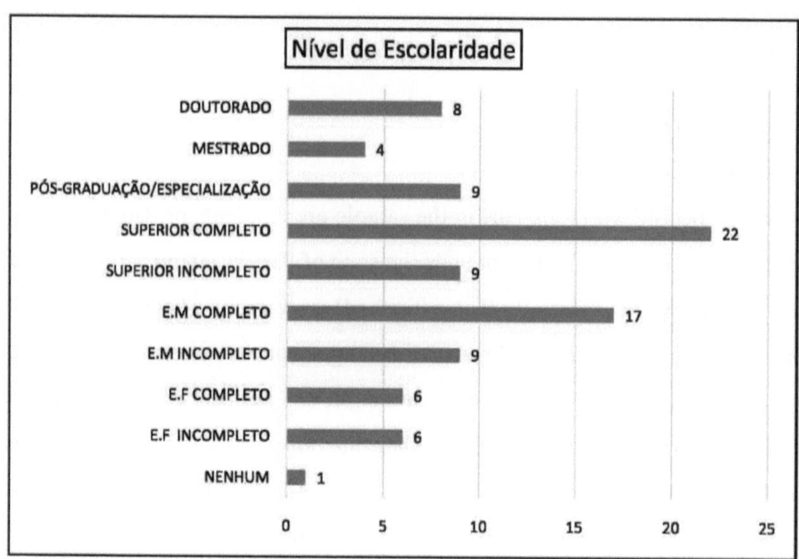

Figure 2- Level of education of the interviewees.

With regard to the relationship with PARNA-Tijuca, it can be seen that most of the interviewees are motivated to go to PARNA-Tijuca for the purpose of sightseeing, which is directly related to leisure in contact with nature (Figure 3). In this question, the interviewee could answer more than one option and include others. The answers obtained reflect the fulfilment of one of the SNUC's objectives, which is set out in article 4, item XII: "to favour conditions and promote environmental education and interpretation, recreation in contact with nature and ecological tourism". It is important to emphasise that even in the minority, some answers should be considered, such as the reconnection[*] [17] which reflects the relationship with nature in the sense of integrating human beings and nature, a question that was raised in the introduction to this research. Physical activity and mountaineering also appear as a response, which indicates that a preserved natural space also stimulates quality of life through sport.

The question about what attracts the most attention was also open to answering more than one option and including others (Figure 4). The aspect that predominates

---

[17]It comes close to a reflection by the philosopher Friedrich Nietzsche: We feel good in the midst of nature because it doesn't judge us.

over all others is the vegetation, which is quite exuberant in many parts, consists of a fragment of the Atlantic Forest biome and is an integral part of the Biosphere Reserve in Rio de Janeiro. Around half of the interviewees were unaware of the history of the Tijuca Forest and that the vegetation there is secondary, i.e. reforested. This data demonstrates the need for environmental education programmes aimed at providing this information to the visiting population and helping to raise awareness. The next most popular options were trails, animals and waterfalls. Thirty-four people (38 per cent) considered PARNA-Tijuca to be quiet, but some people disagreed. They thought it was too noisy, and attributed this mainly to two factors: the behaviour of groups walking through the park. And also because of the number of vehicles on both the internal and external roads. This issue is most evident in Parque Lage, especially in areas close to the entrance, where it borders Rua Jardim Botânico, which has an intense flow of cars. This characteristic is inherent to an urban forest, but it is still possible to think of alternatives to reduce the noise, or excess noise.

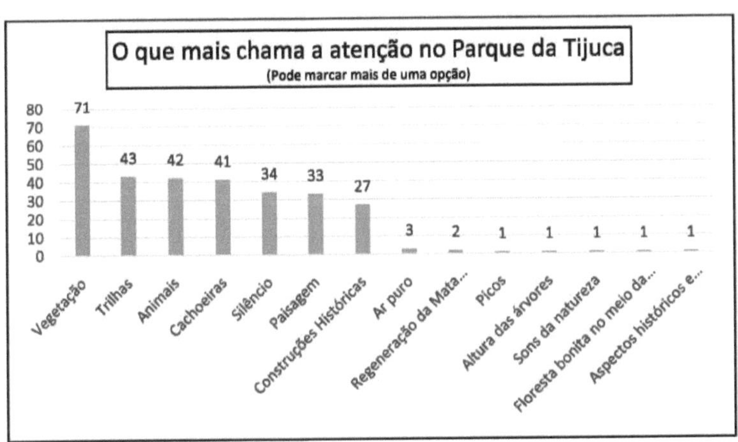

Figure 4 - what attracts the most attention at PARNA-Tijuca.

The responses regarding the importance of PARNA-Tijuca were very interesting (Figure 5). This question was free, there were no suggested answer options, and there was no limit to the number of answers; people could attribute importance to PARNA-Tijuca for various reasons. The answers were grouped together because they had the same meaning, and others were considered in isolation. People judged PARNA-Tijuca to be important mainly because of its environmental preservation and the "clean air" it provides, as it doesn't concentrate so much pollution. The metaphor "lung of the city" was also used for similar answers[18] .

The following answers were given: leisure, acclimatisation of the city (lowers the temperature), tourism, water supply, source of more than 40 rivers, erosion control, protection of fauna, remnant of Atlantic forest, general environmental services, reconnection of man with him, etc.

The park is also important for the environment, capturing noise and atmospheric pollutants, retaining rainwater, controlling river silting, collaborating with the city, endangered endemic species, environmental education, improving quality of life, organising the city and helping to get away from the routine of urbanisation. Four interviewees answered very important, without justifying its importance, and eighteen people didn't answer, which corresponds to 20% of the sample. The answers seem quite coherent, in line with the objectives of a conservation unit in the National Park category and in line with conservationist ideas.

According to Article 11 of the SNUC/2000, the objectives of the National Park are:

The basic objective of the National Park is to preserve natural ecosystems of great ecological importance and scenic beauty, enabling scientific research and the development of environmental education and interpretation activities, recreation in contact with nature and ecological tourism.

---

[18]This metaphor is misleading: the forest as the "lung of the city" or the "lung of the world" is better applied to the oceans, where the production of oxygen by photosynthesising algae is predominant due to the greater extent of the oceans than the forests. The most appropriate metaphor for the forest would be the "refrigerator of the city" or the "refrigerator of the world", because it makes the climate milder due to the humidity it provides.

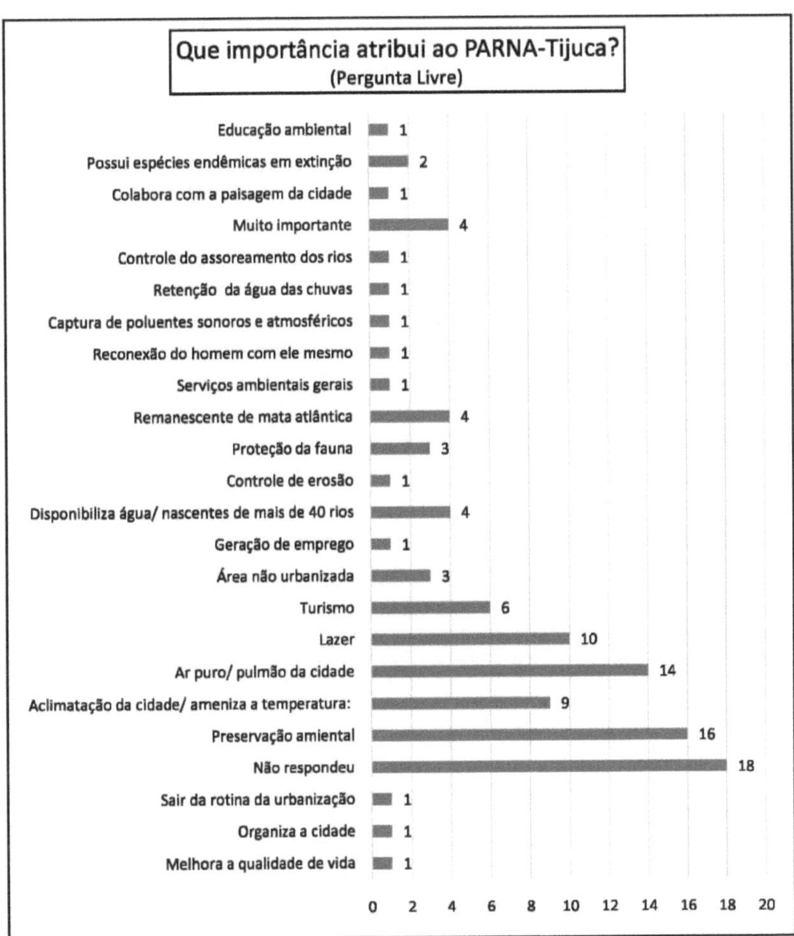

Figure 5 - importance attributed to PARNA-Tijuca.

With the aim of understanding the knowledge of the population that frequents a Conservation Unit and the population living in the surrounding area about concepts related to the environment, the questionnaire included some expressions that are used in academic circles and also in the media. They had to answer yes or no as to their knowledge of each expression, without explaining in detail what it was. In the case of From the direct interviews, i.e. those that were not conducted over the internet or the questionnaire delivered later, it was possible to exchange more information and make some clarifications about the concepts.

43

Specially Protected Area[19] (Figure 8) is a concept known by more than half of the sample. The name Conservation Unit[20] (Figure 7) was also familiar to more than half of the sample. The concept of Urban Forest[21] (Figure 9) was familiar to 67 per cent of the sample, which corresponds to sixty people.

The Buffer Zone[22][23] (Figure 10) is a concept known to more than 50 per cent of the sample. The term environmental impact is also well known: 67 per cent of those interviewed were familiar with the term, which is easy to deduce.

The concept of the National System of Conservation Units (SNUC)23 (Figure 6), on the other hand, is unknown to fifty-four people in the sample, which corresponds to 60 per cent. It should be considered that this is a very technical term that is little talked about outside of academia^.

The concepts of the Forest Code[24] (Figure 12) and the Environmental Crimes

---

[19]Protected areas are territorially demarcated spaces whose main function is the conservation and/or preservation of the natural and/or cultural resources associated with them. According to the World Conservation Union (IUCN), they can be defined as "a terrestrial and/or marine area specially dedicated to the protection and maintenance of biological diversity and associated natural and cultural resources, managed through legal or other effective instruments" (IUCN, 1994:7).

[20]A Conservation Unit (CU) is a territorial space with relevant natural characteristics and defined boundaries, established by the Government to guarantee the protection and conservation of these natural characteristics/ MMA/

[21]Two concepts have been used in Brazil to designate all the tree vegetation present in cities: Urban Afforestation and Urban Forest. Both have had their content redefined recently, probably based on the terms established by Canadians and Americans from the 1960s onwards. The history of the concept of "Urban Forest" is linked to the expansion of cities and the growing demand for methods and techniques that could be applied to the arboreal ensemble of these spaces. Grey & Reneke (1986) explain that this definition first appeared in Canada, quoted by Erik Jorgensen (1970), who described the Urban Forest as the set of all the trees in the city, present in the streets, watersheds, recreational areas, their interfaces and spaces of influence.

[22]The surroundings of a conservation unit, where human activities are subject to specific rules and restrictions, with the aim of minimising negative impacts on the unit (SNUC 2000).

[23]The National System of Conservation Units (SNUC) is the set of federal, state and municipal conservation units (UC). It is made up of 12 categories of CU, whose specific objectives differ in terms of the form of protection and permitted uses: those that require greater care, due to their fragility and particularities, and those that can be used sustainably and conserved at the same time. MMA/2000.

[24]The New Brazilian Forest Code (Law No. 12,651, of 25 May 2012, originating from Bill No. 1,876/99) is the Brazilian law that provides for the protection of native vegetation, having repealed the 1965 Brazilian Forest Code. Since the 1990s, the proposal to reform the Forest Code has sparked controversy between ruralists and environmentalists. The bill that resulted in the current text spent 12 years in the Chamber of Deputies and was drafted by deputy Sérgio Carvalho (PSDB from Rondônia). In 2009, deputy Aldo Rebelo of the PCdoB was appointed rapporteur of the bill and

Act (Figure 13) seem to be well known. Both concepts received a response of around 56 per cent. This can be attributed to the controversies generated among some social sectors during the voting process by the National Congress on the New Brazilian Forest Code in 2012. As for the environmental crimes law, people made associations with the punishment of offences that threaten the environment, in situations where there are seizures and arrests, such as biopiracy, the arrest of groups of balloonists, the punishment of oil exploration companies in the aquatic environment, among others.

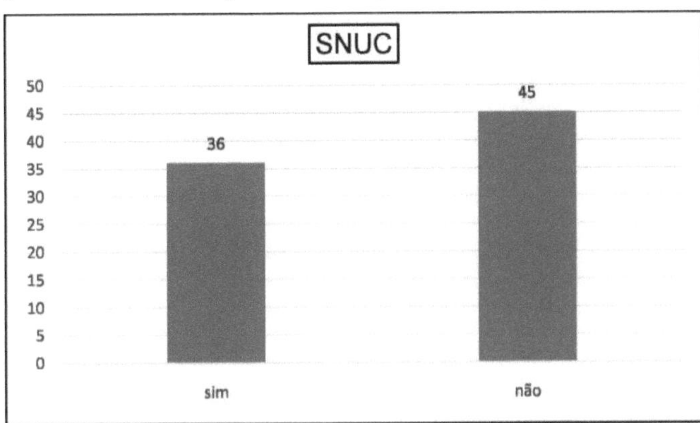

Figure 6 - Knowledge of the SNUC.

issued a report in favour of the law in 2010. The Chamber of Deputies approved the bill for the first time on 25 May 2011, sending it to the Senate. On 6 December 2011, the Senate approved Aldo Rebelo's bill by 59 votes to 7 (in the Senate, the bill was renamed "House Bill 30 of 2011"). On 25 April 2012, the House approved an amended version. In May 2012, President Dilma Rousseff vetoed 12 points of the law and proposed changes to 32 other articles. After Congress approved the "New Forest Code", NGOs, activists and social movements organised the "Veta Dilma" movement, calling for a full veto of the bill) and the Environmental Crimes Law (footnote: The Environmental Crimes Law is a Brazilian law. Law No. 9.605, of 12 February 1998. Sanctioned by then President Fernando Henrique Cardoso, it provides for criminal and administrative sanctions arising from conduct and activities harmful to the environment.

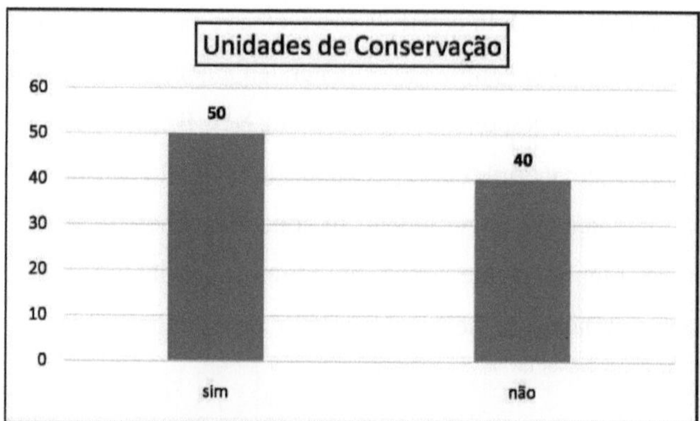

Figure 7 - knowledge of Conservation Units.

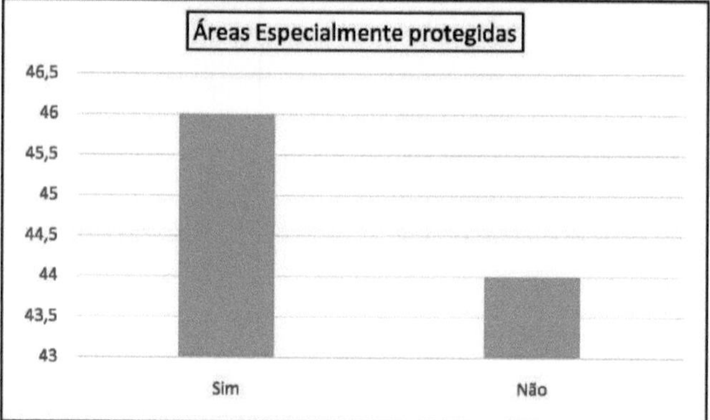

Figure 8 - Specially Protected Areas.

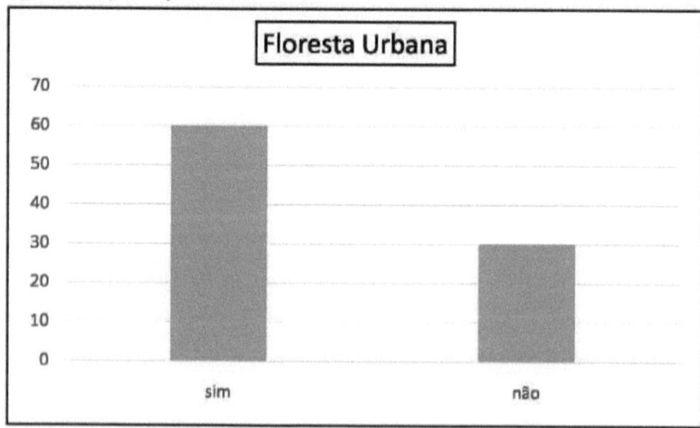

Figure 9 - urban forest.

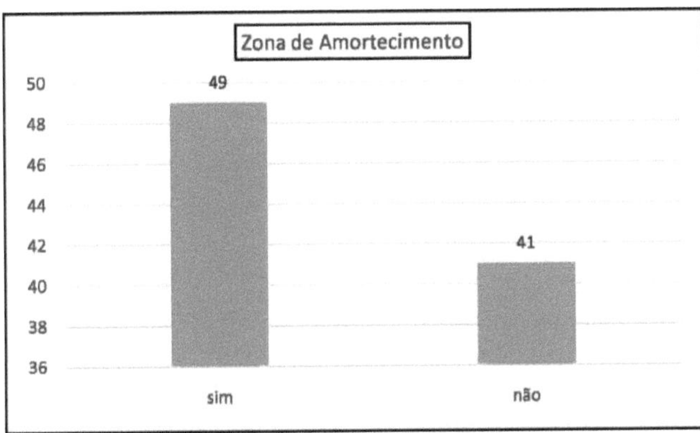

Figure 10 - buffer zone.

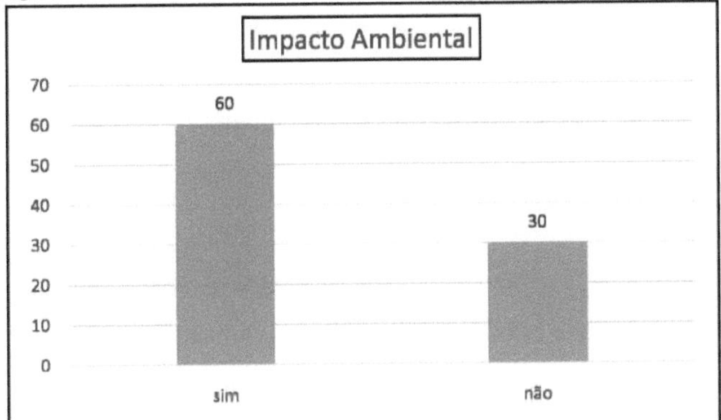

Figure 11 - Environmental impacts.

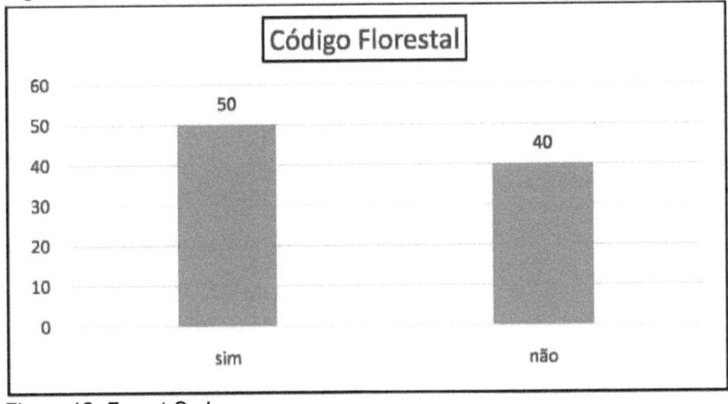

Figure 12- Forest Code.

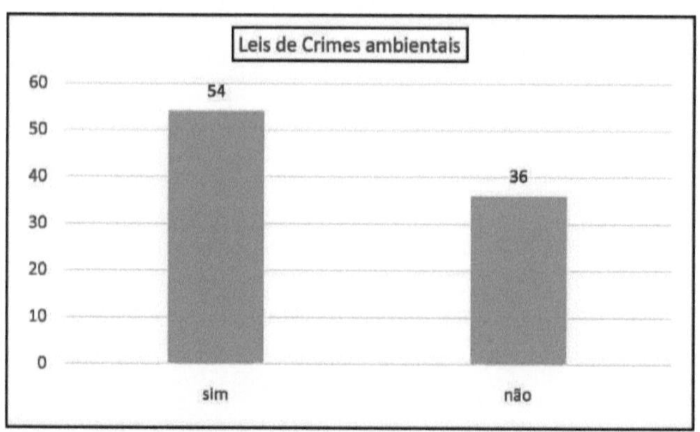

Figure 13- Environmental offences law.

The majority of those interviewed (67 people, corresponding to 68 per cent) believe that impacts outside the core area of PARNA-Tijuca interfere with the core area. There is a consensus that the "edge effect" jeopardises the core area of a conservation unit. Edge effects are the ecological results of physical and biological alterations at the contacts of the forest fragment (LAURANCE, 1991). In the Atlantic Rainforest, edges are the result of natural processes such as the emergence of clearings or natural geographical limitations such as rocky outcrops and riverbanks. However, it is the edges created by man that are causing great concern. That's why all Conservation Units, with the exception of Private Natural Heritage Reserves (RPPN) and Environmental Protection Areas (APA), both included in the sustainable use category, according to the SNUC must have a buffer zone, with boundaries defined in the Conservation Unit's Management Plan. The purpose of the buffer zone is to prevent external pressures from interfering with the core area, i.e. to minimise negative impacts on the unit (SNUC/2000). The borders mentioned by Laurance can be found in the buffer zone. According to the results in Figure 14, twenty people (22%) believe that the impacts on the external area do not interfere with the core area and three people (4%) were unable to answer, showing doubts about this process.

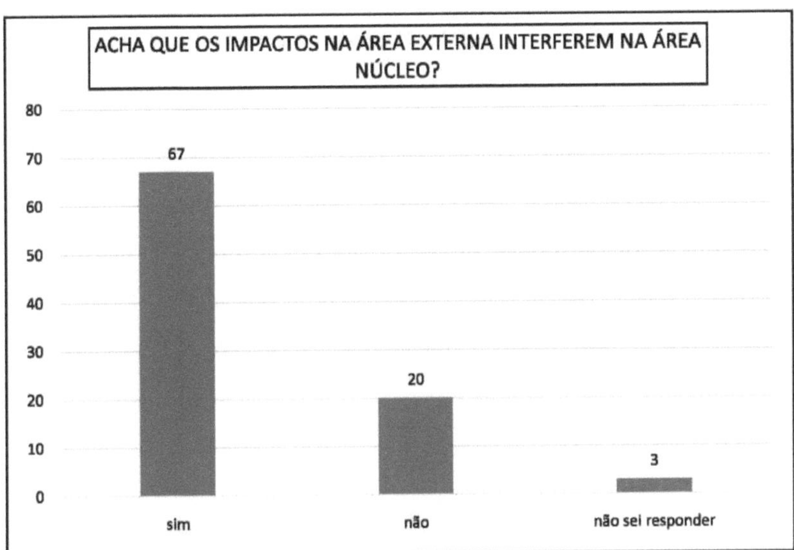

Figure 14 - knowledge that impacts on the external area interfere with the core area.

Through the answers to the questionnaire and direct observations, it was possible to see that in the area outside PARNA-Tijuca there are no mechanisms, either from the public authorities or from civil society, to provide adequate environmental guidance, both for the population living in the neighbourhood and for visitors to PARNA-Tijuca. Some people in the sample who answered yes to the guidance they received outside the park considered it to be training that takes place through the PARNA-Tijuca environmental education team (Figure 15).

In the surroundings of the four sectors, visual information about the biological value of the area and directions to the entrances to PARNA-Tijuca was poor. The current signposts are precarious. But the situation is worse in sector C, along Estrada das Canoas. There are no signs indicating the ascents to Pedra da Gávea and Pedra Bonita. This prevents and hinders access for visitors. In addition, efficient guidance in the surroundings is also a way of protecting the core area, and it is important for citizens to situate themselves and realise that they are passing through a biodiversity conservation site. It's important not only for the knowledge itself, but also for the type of appropriate behaviour that aims to preserve the area. Guidelines should be given by the environmental agency responsible for PARNA-Tijuca and included in the Management Plan. For example, care should be taken when crossing the road with wild animals, when dumping waste, especially cigarette butts, among others.

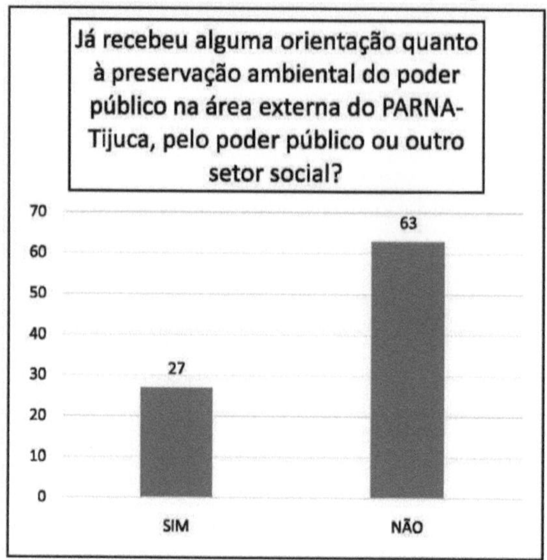

Figure 15 - guidance received in the outdoor area.

The majority of the people who answered the questionnaire with the greatest accuracy are part of the group of PARNA-Tijuca employees. This is a group that demonstrates good training, knowledge and involvement with PARNA-Tijuca. According to the PARNA-Tijuca Environmental Analyst, in an interview for this research:

[...] the park offers training courses. It is developed by the environmental education team and given periodically to new employees. Some employees not only absorb the course content well, but also become partners with the park, helping with various conservation tasks. Perhaps it is easier for them to perceive anthropogenic impacts on the park's biota on a day-to-day basis, something that hardly ever occurs to visitors.

5.1- A SURVEY OF THE MAIN ENVIRONMENTAL IMPACTS IN AND AROUND THE TIJUCA NATIONAL PARK:

With regard to the impacts observed in the external area, the three most pointed out impacts were: vehicle pollution, river pollution and rubbish. These were followed by: fires, the presence of exotic fauna and flora, hunting, transmission masts, irregular occupation (favelas), noise, festivals, religious offerings, deforestation and general population occupation. It is important to note that eleven of the ninety people in the sample (12 per cent) said that they did not observe any impacts on the area outside PARNA-Tijuca. We can deduce some justifications for this way of looking at the surroundings. One is that this portion of the sample really wasn't aware of the environmental degradation in the area outside PARNA-Tijuca. And the other explanation is that even though people observe the degradation, they don't consider it as such, it's as if the impacts on the environment were part of that landscape (Figure 16). This fits in well with what Ferrara (1999) says: environmental language and the perception that users of a place have of it have their existence identified by the observation that captures and records the images and associates them. On the other hand, the rapid transformation that is the sign par excellence of the modern city relativises those images in a short space of time.

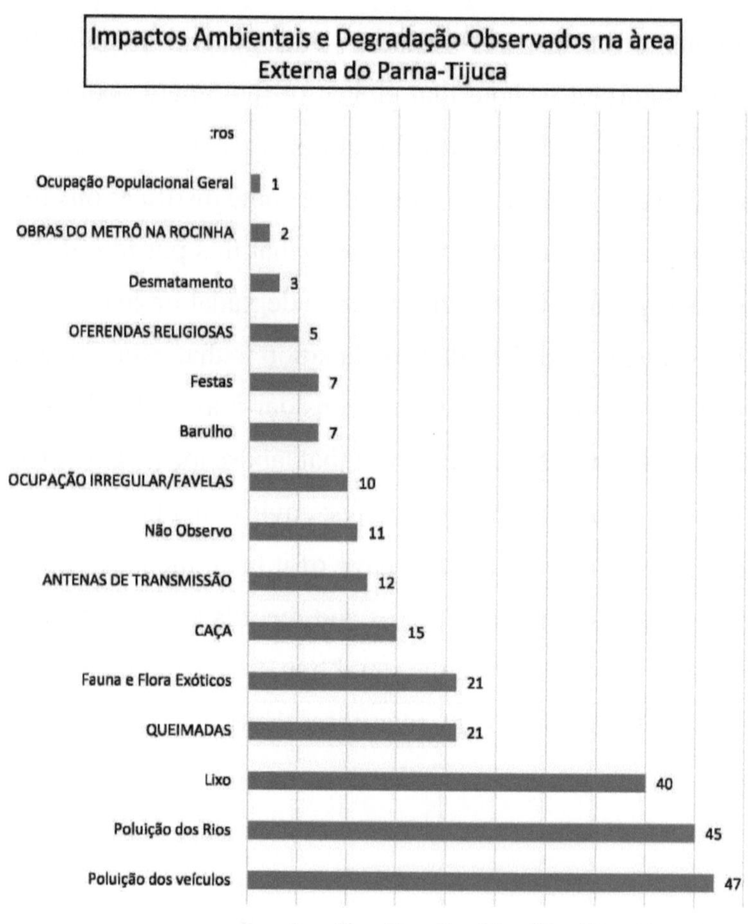

Figure 16 - environmental impacts pointed out in the area outside PARNA-Tijuca.

5.1.1 - Vehicle pollution

It can be inferred that because PARNA Tijuca is located in the central region of the city of Rio do Janeiro, it is cut off by the road network that connects the northern, southern and western regions of the city. A large number of motor vehicles pass through the park, emitting pollutants through the combustion of petroleum products on a daily basis. The effects of the local concentration of pollutants and their effects on the atmosphere in the Park and surrounding areas should be assessed in terms of physical, chemical and biological aspects, so that any damage can be measured in the future (PLANO DE MANEJO/ PARNA- Tijuca, 2008).

The concentration of pollutants in the air is the result of emissions from stationary sources (industries, incinerators, etc.) and mobile sources (motor vehicles) combined with other factors such as climate, geography, land use, distribution and typology of sources, emission conditions and local dispersion of pollutants (IBGE, 2014).

The Metropolitan Region of Rio de Janeiro (RMRJ) has a high concentration of pollution emission sources that generate serious problems. Of the country's metropolitan regions, this is the most densely populated, with approximately 1,900 inhabitants per square kilometre, spread over 19 municipalities that occupy 14.9% of the state's total area. In an area of just under 6,500 km2, they concentrate a population of eleven million people, around 80% of the entire state, 60% of whom live in the municipality of Rio de Janeiro. This region is home to the second largest concentration of vehicles, industries and sources of air pollution in the country. It also has physical characteristics that increase the problems related to air quality: the rugged topography of the region; the presence of the sea and Guanabara Bay, which together produce a complex and heterogeneous air flow in terms of the distribution and dispersion of pollutants; and the tropical climate, which favours photochemical processes and other reactions in the atmosphere, generating secondary pollutants. Added to these physical factors is the heterogeneous and intense occupation of the land (PLANO DE MANEJO/ PARNA- Tijuca, 2008).

Air pollution in large urban centres is one of today's major environmental problems, with serious implications for the health of the population, especially children, the elderly and those suffering from respiratory diseases such as asthma and respiratory failure. From this point of view, while the annual maximum values highlight critical pollution events and moments ("acute pollution"), the annual averages show the common, normal state of the atmosphere, highlighting what we can call "chronic pollution". Because of this, the critical values of the CONAMA standard for annual average concentrations are much lower than those for daily values (IBGE/2014).

PARNA-Tijuca is most influenced by Avenida Brasil and Avenida das Américas, which are the roads responsible for most of the pollutant emissions. Due to the immense flow of vehicles, Avenida Brasil is responsible for 25 to 30 per cent of the total air pollutants emitted by traffic routes in the Rio de Janeiro Metropolitan Region (PLANO DE MANEJO/ PARNA- Tijuca, 2008).

The interaction observed between biotic and abiotic factors is reflected in the fragility of the vegetation that covers the Tijuca massif. This means that external changes associated with anthropogenic processes can alter the functioning of the internal operations that regulate the dynamics of the forest geoecosystem (COELHO NETTO, 1990). Living in the metropolis of Rio de Janeiro leads to continuous exposure to chemical substances from industrial waste, means of transport and terrestrial dust.

The atmosphere contains elements from natural sources such as the oceans and terrestrial ecosystems, including sodium (Na); potassium (K); magnesium (Mg); chlorine (Cl); hydrogen (H); iron (Fe); manganese (Mg); zinc (Zn); chromium (Cr); copper (Cu); lead (Pb), cadmium (Cd); and nickel (Ni). Urban-industrial centres contribute to the increase in substances injected into the atmosphere. Of particular note are nitrogen and sulphur oxides resulting from the burning of fossil fuels (coal and oil) and high-temperature steelmaking processes: these substances become acidic when in contact with water (COELHO NETTO, 1990).

The lead content, which also comes from fossil fuels, is at risk of increasing beyond the levels tolerated by the World Health Organisation (0.5 to 1 microgram/m3), not only because of the high concentration of motor vehicles, but also because of industrial sources. The precipitation of acidic water and lead on forests can alter the relationships between geo-environmental factors, causing toxic effects that degenerate the biota. However, the impact of acid rain can be controlled by the neutralisation capacity of the affected environment (COELHO NETTO, 1990).

5.1.2 -River pollution

The Tijuca massif is drained by several articulated canal systems, which receive water and sediment flows from the slopes during rainy periods. These systems converge on the artificial canals that drain the city in the lower areas. In the Serra da Carioca, the following canals stand out as the main collectors: the Trapicheiro and Comprido rivers, which rise on the NW slope and descend to the north of the city towards the Mangue canal; the Carioca river, which rises to the SE of Paineiras, receives the Silvestre and Lagoinha rivers as tributaries,

and crossing the neighbourhood of Laranjeiras flows into Guanabara Bay; and the Cabeça, Rainha and Macaco rivers which flow into the Rodrigo de Freitas lagoon. Draining the Tijuca mountains are the Joana and Maracanã rivers, which also converge into the Mangue canal; on the southern slope, the Cahoeira river flows into the Tijuca lagoon, and towards Jacarepaguá, the Anil and Panela rivers flow (COELHO NETTO, 1992).

Observing the appearance of the water flowing down the slopes and some of the waterfalls around PARNA-Tijuca clearly shows pollution. In some places (Furnas de Agassiz, for example), you can see solid waste adhered to the bottom and in suspension and a typical sewage smell, which indicates the discharge of sewage. Foam also forms at some points. These observations indicate that the water quality is affected and possibly not in line with environmental legislation.

Figure 17 - Furna de Agassiz waterfall, both situations with a high level of pollution.

All these springs have suffered environmental impacts in different ways throughout the formation of the city of Rio de Janeiro. This has occurred and is occurring through the discharge of "in natura" sewage, industrial sewage or irregular abstraction. Sewage is discharged predominantly from diffuse sources[25] , as indicated by the observations and data from the questionnaire survey, which reflects the way in which the city was colonised and occupied.

Sanitary sewage systems, both in terms of quantity and quality[26] are still a major challenge for the country's public policies, since their cost is recognised as high and they require appropriate planning, design and construction technologies, due to the multiple factors involved. With the growth of cities and water consumption, this issue is getting worse every day (DIAS, 2003).

According to the IBGE (Brazilian Institute of Geography and Statistics), 71.8

---

[25]Diffuse pollution is defined as the discharge of polluted water into the water network that runs through the urban area from sources that are not clearly identified.

[26]Water quality is represented by a set of generally measurable characteristics of a chemical, physical and biological nature. As it is a resource common to all, it has been necessary to establish legal restrictions on its use in order to protect bodies of water. Thus, the physical and chemical characteristics of water must be kept within certain limits, which are represented by standards, guiding values for the quality of water, sediment and biota (Conama Resolutions No. 357/2005, Conama No. 274, Conama No. 344/2004, and Ordinance No. 518 of the Ministry of Health).

per cent of municipalities did not have a municipal basic sanitation policy in 2011. The statistic corresponds to 3,995 cities that do not respect the National Basic Sanitation Law, approved in 2007.

The majority (60.5%) had no monitoring of sanitation licences, in addition to drainage and management of urban rainwater and water supply. In almost half of the country's cities (47.8 per cent), there is no water quality inspection body (IBGE, 2011).

The evolution of sanitation systems is linked to the institutional development of the state, the mode of production, technological development and income distribution. The spread of poverty throughout the urban fabric hinders the preservation of natural resources and artificial environmental protection systems, so that poverty can be considered the greatest obstacle to the sustainability of ecosystems. Developing environmental sanitation requires, first and foremost, socio-economic and political solutions relating to employment and income (DIAS, 2003).

The city of Rio de Janeiro belongs to the state's western sewage sub-basin, which is made up of three systems according to its topography: Guanabara, where effluent is sent to Guanabara Bay and the Atlantic Ocean; Jacarepaguá, where effluent is sent to the Atlantic Ocean; and Sepetiba, which discharges sewage into Sepetiba Bay (Dias, 2003).

The city has a water system that includes around 250 rivers and canals and complex lagoon systems, including the Rodrigo de Freitas Lagoon and the lagoons of the Baixada de Jacarepaguá. These are the receiving bodies for rivers and streams from the Tijuca and Pedra Branca massifs to the ocean. Its long coastline (86 kilometres), bordered to the east by Guanabara Bay, to the west by Sepetiba Bay and to the south by the Atlantic Ocean, has 72 beaches (PREFEITURA, 2001, v.9).

The rivers of the city of Rio de Janeiro are characterised by their modest volume of water, the sinuosity of their courses, the absence of a dominant direction and the difficulties of flow due to the extensive flat areas with low elevations. These characteristics, combined with a tropical rainfall regime with intense summer rains, make the rivers susceptible to overflows, periodic flooding and permanent flooding in certain areas, intensified during periods of rising tide levels. In the dry season, their flow is considerably reduced. Most of them are channelled, straightened and covered by streets and avenues (PREFEITURA, 2001, v.9).

Among the city's rivers, the Carioca River, as the first surface water source, has played a strategic role for the city for a long time and is a natural landmark of the city. Even so, this internationally recognised body of water, an affective reference for the people of Rio de Janeiro, is seriously compromised in its environmental aspects. Its recent history is still marked by precarious and discontinuous sanitation interventions, of a questionable conception and provisional nature, which perpetuate the problems arising from the lack of control, monitoring and inspection of the sources of pollution (PESSOA, 2010).

Rio de Janeiro has spent its entire history making great efforts to fight for water. As Coaracy (1965) puts it, "since its origins, Rio de Janeiro has always been a thirsty city". Various expansion projects have been implemented in an attempt to solve different problems related to the components of the public water supply system - collection, supply, treatment, reservoir and distribution (COROACY,1965).

Faced with growing demands, associated with the advance of the deforestation process, pollution and due to their low supply capacity, these water collections gradually ceased to serve as springs. Even during the imperial government, in 1870, the solution adopted was to collect water from increasingly distant springs (TELLES, 1984).

From then on, the old springs, already compromised by the pollution process, were only remembered when floods hit the city. The city is made up of areas of great ecological vulnerability, due to their susceptibility to erosion and flooding. Urban sprawl increases the risks, as it triggers systematic environmental degradation with continued land consolidation, deforestation and asphalting, which aggravate the great structural fragility of its natural environment (PREFEITURA, 2001, v.9).

Sewage channels or ditches have been the common name adopted by the population of Rio de Janeiro to designate rivers, canals and streams that run through their city, and which should be recognised as such. In view of the evidence, it is difficult for any specialist to undo this confusion and discriminate between the causes that have culminated in the current state of environmental degradation (DIAS, 2003). High density urbanisation inevitably generates environmental impacts on natural resources. However, the impacts on river basins are the result of the inadequate use of their resources rather than the exploitation of the natural wealth available there.

The way in which the city was metropolised - without urban planning or concern for ecosystems - has increased the negative environmental impacts on river basins. The main negative actions on water assets are: the discharge of sanitary sewage; the discharge of non-domestic effluents (industrial, hospital, petrol stations, etc.); the discharge of fertilisers and pesticides from the river.); dumping of fertilisers and pesticides from agricultural activities; sealing of the soil; dumping of solid waste, including floating waste; carrying sediment, nutrients, humus from the soil, oil derivatives and solid waste into the river channel through surface run-off; landfills, both authorised and illegal; subtraction of the riverbed and floodplain areas for incorporation into urban functions; de-characterisation of natural conditions through diversions and channelling of the riverbed; disorderly occupation of the banks; deforestation; silting up; sand extraction; atmospheric pollution; pollution resulting from environmental accidents, etc. (DIAS, 2003).

It is important to emphasise that within the Conservation Unit surveyed, the way water is managed is also inappropriate. Some information in this regard was widely pointed out during the questionnaires. One of the findings is that there is no sewage treatment system to treat the effluents generated at PARNA-Tijuca's facilities, which in turn are discharged into the park's own water bodies. Another is the abstraction of water by CEDAE, which is seen by common sense as harmless. The abstraction of water from springs is a much more serious problem than it first appears. Both legal abstraction by the State Water and Sewage Company (CEDAE) and irregular abstraction by local residents are harmful to the Tijuca Forest.

If, on the one hand, it is known that most of Rio de Janeiro's water supply comes from pipelines that connect to the Guandu Lameirão system, in the Serra do Mar, it is important to remember that the Tijuca Forest (and the Pedra Branca Massif) still have their various streams and rivers dammed by CEDAE's catchments. The Tijuca Forest was created for the purpose of forming water sources; after all, its reforestation, which began in 1861, was primarily aimed at restoring the forests at the headwaters of the rivers, in order to put an end to the recurring droughts that plagued Rio de Janeiro at the time (O ECO, 2004).

In order to protect water bodies and maintain their natural characteristics, it is essential to delimit their protection zones in order to regulate activities in them (MOTA, 1995). However, these areas, from which large tracts of land are taken, are under constant anthropogenic pressure for various purposes, such as the expansion of the road system and the construction of irregular housing, which will unduly discharge their effluents into water bodies. The pollution of water bodies by solid and liquid waste results in the release of gases from these effluents, as well as those generated by the process of anaerobic decomposition, resulting in atmospheric pollution.

According to the Master Plan - Environment (PCRJ, 2000), the city's sewage situation is chaotic. The toxic parameters of chemical pollution from industries that discharge effluents into water bodies are monitored by FEEMA in twenty-five rivers, of which nine are saturated, thirteen are in the process of being saturated and only three are not yet saturated.

The prospect of controlling diffuse pollution seems distant in a scenario where even point source pollution[27] has not been controlled in practice. Therefore, the concentration of efforts and resources indicates the need to prioritise the control of industrial and domestic point source discharges in Brazil. Diffuse pollution and its effects cannot be disregarded, and their equation must be articulated with point sources, since the causal relationships between these problems are interrelated and interdependent (DIAS, 2003).

Problems such as flooding and water pollution are the result of inadequate management of natural resources, in which the process of occupying space prioritises political and economic interests, degrading the environment and reducing the quality of life (CUNHA, 1996). Therefore, not only must scientific studies be carried out in

---

27Where the source of pollution is easily identifiable as an emitter of pollutants.

these environments, but society must also participate and raise awareness of the rational use of resources, so that nature's response to anthropic actions does not cause even greater damage to the population and the ecosystems that still exist.

5.1.4 - Rubbish, solid waste

The issue of solid waste is a problem in most areas where there is any level of human occupation. And in Brazil, environmental legislation in this regard is fairly recent. As such, the poor management of solid waste in Brazil and in the city of Rio de Janeiro is also reflected in most of the national territory, including the core area and the Tijuca National Park's neighbourhood. Below are photos, one in the core area (sector C, Pedra Bonita), in which we can see an attempt at selective collection, but the rubbish bins are deteriorated, and another in the external area (sector A, Alto da Boa Vista) in which rubbish has accumulated in the bed of a spring.

Just over 32% of the country's municipalities (1,796) have an active selective waste collection programme, project or action, according to an IBGE survey. On the other hand, 2,376 cities (42.7 per cent) still don't have any kind of selective waste collection initiative. 3.3 per cent of the cities have a pilot selective waste collection project, but only in restricted areas. Meanwhile, 2.5% of cities have even started such programmes, but have stopped for unspecified reasons. In terms of urban cleaning services, the South stands out in the study on the profile of Brazilian municipalities, with 663 cities that have selective collection, which represents 55.8 per cent of the rest of the country. The Southeast, with 41.5 per cent (693 cities), is second in the regional ranking (IBGE, 2011).

On the other hand, the North and Northeast regions have the highest proportions of municipalities without programmes, 62.8% (282) and 62.3% (1,118), respectively. According to the IBGE, selective waste collection is more common in large cities: 68.2% (193) of municipalities with more than 100,000 inhabitants report having an active programme (IBGE, 2011).

High consumption rates and growing waste production are among the biggest environmental problems facing humanity. It is obvious that the problem is exacerbated by the expansion and densification of urban agglomerations, since the sanitation infrastructure in most Brazilian cities does not keep pace with this rapid growth.

Law 12.305 of 2 August 2010, which establishes the National Solid Waste Policy, lists many issues, and its analysis is not a one-off. Waste production intensified with the industrial revolution, growing more and more over the years, and today, with a consolidated consumer society, it is an urgent issue facing humanity. Policies are therefore needed to reverse the chaotic situation caused by the excessive generation of waste in all parts of the world. This issue is highly relevant for two main reasons: the great exploitation of exhaustible natural resources, and the destination of waste in terms of storage, which according to current data is in places of greatest poverty (ALIER, 2007).

In Brazil, the first legislative initiatives focussing on solid waste emerged more than twenty years ago, in the 80s. Hundreds of related projects have been processed together. These bills were analysed by special committees and some were considered unconstitutional. In 2008, a Working Group was set up in the Chamber of Deputies to make it possible for the National Congress to deliberate on the matter, and it was sanctioned by the Presidency of the Republic in the form of Law No. 12.305 of 2 August 2010.

This law proposes many changes in Brazilian society, even though it is expected to be implemented in another two decades. There is a dilemma in finding measures to make integrated solid waste management viable, as it depends on the efforts of individual citizens, social spaces (families, educational institutions, businesses, among others), and the commitment and political will of public authorities.

5.1.5 - Burning

Studies indicate that fires linked to natural phenomena have been burning in the Tijuca massif since ancient times. It is not possible to diagnose the frequency or extent of fires in this period. But it can be assumed that it was more linked to natural phenomena, since there was no human occupation in the region during this period.

The increasing occurrence of burn-offs and forest fires in Brazil at the end of the 18th century began to mobilise enlightened sections of Brazilian society against the destruction of natural resources (PÁDUA, 2002). Forest fires and the practice of burning played an important role in the historical occupation of the current area of PARNA Tijuca.

Brazil, like other countries, has been studying measures to prevent and combat fire in its protected areas since the beginning of their creation (CAREY et al, 2000). This type of impact occurs mainly in the buffer zone, which threatens the core area of the protected areas.

Currently, one of the activities of IBAMA's National System for Preventing and Fighting Forest Fires (PREVFOGO) is to provide technical and financial support for preventing and fighting forest fires in various federal Conservation Units, one of which is the Park. Since 2000, PREVFOGO's activities have been strengthened throughout the country.

With the hiring of temporary brigades, it is estimated that there has been a 50% reduction in fires in the conservation units managed by IBAMA (RAMOS E BOSNICH, 2005). Despite the existence of records since 1979, one of the frequent failures in the management of forest fires in the units managed by IBAMA, now the Chico Mendes Institute for Biodiversity Conservation (ICMBio), is obtaining fire occurrence records (ROI), systematising the data and consequently transferring it internally and disseminating it externally (MORAIS, 2004).

In order to characterise the history of fire in PARNA Tijuca, we will use the Park's archived records and, in addition, we will use two studies from 2006, which have statistics on the occurrence of fire in the area (SILVA 2006).

Analysing this data is an essential tool for planning prevention and firefighting strategies, as it points to critical areas and periods, as well as providing support for specific protection actions, making it possible to determine the causal agents. In addition, this data is indispensable for validating models of the behaviour of advancing fire fronts and fire risk indices (CARAPIÁ, 2006).

Between 1997 and 2006, records show that there were 20 outbreaks of fire in the Park and surrounding area, which consumed 127 ha and 69 ha of vegetation, respectively. Of the total area burnt (196 ha), 45% occurred in forest vegetation (87 ha in the interior and 2 ha in the surrounding area) and 55% in areas with altered vegetation (40ha in the Park and 67ha in the surrounding area). Within the Park, there were relatively large fires (>10 ha) in all three sectors (A, B and C), but most of them (67%) reached an area of no more than 5 ha. However, the number and extent of fires in Atlantic forest in the middle and mature stages of succession (6 ha) were relatively small. When a forest is subjected to fire, the changes it produces in the quality, quantity and continuity (vertical and horizontal) of fuels in the vegetation increases its likelihood of recurrence and its future intensity.

According to Matos (2002), the annual distribution of fires in the Tijuca Massif is asymmetrical, with years of relatively high frequency with more than 100 outbreaks (1993, 1995, 1997), interspersed with years in which there were between 40 and 100 records (1992, 1994, 1998, 1999) and years of low occurrence, with less than 20 outbreaks (1991, 1996, 2000).

Each year, fire outbreaks are more frequent during the dry winter months (June to August), although events have been recorded in all seasons. During the summer (January to March), a relatively high number of cases are observed in the late afternoon, reflecting periods of drought (over 15 days) which, together with the high temperatures of this season, create a favourable framework (veranicos) for the occurrence of fires (CARAPIÁ, 2006).

According to the authors, although roads facilitate access for firefighters, they are also the main vector for spreading fires and facilitate the invasion of exotic plant species (Panicummaximum and Pteridiumaquilinum) that make the vegetation more flammable. These two species are frequent in the seed bank of areas with a history of fires (BELINATO E SILVA MATOS, 2003).

The park's records indicate the existence of some critical areas inside the park and in the surrounding areas. In Sector A, 65 per cent of the occurrences and 78 per cent of the burnt area were recorded, with two locations standing out: Morro do Elefante (6 cases) and Pedra do Conde (3 cases). The ridge line of Morro do Elefante, which relatively overlaps the perimeter of the Park, shows a sudden change in forest vegetation (towards its interior) with an extensive degraded area adjacent to the UC where grasses predominate. There, the degraded area totalled around a third of the total area burnt in the Park and its fires were caused by crime. Probably caused by the burning of leftover agricultural crops or the burning of plant material for pasture renovation (PLANO DE MANEJO PARNA-TIJUCA, 2008).

In sector B, three fires were recorded as a result of balloons and religious offerings, which resulted in the burning of around 27 ha. Of particular note here is the fire that occurred in 1997 in the Trapicheiros River basin (north side of Sumaré), which was relatively large (25 ha), and which resulted in a canopy fire, according to reports from park staff. In this sector, the biggest concern is the intense and disorganised growth of communities or favelas, which were previously sparse and much smaller in number (PARNA-TIJUCA MANAGEMENT PLAN, 2008).

In this case, the factors that most contribute to the current occurrence of forest fires are the greater deposition of pollutants by this population and the greater exposure to the sun on this northern slope of the massif, due to forest loss (COELHO NETTO, 2005). For the author, another cause of forest devastation is the installation of television towers in Sumaré, which contributes to the destabilisation of the slopes in the Park, and increases the favourable situation for the occurrence of forest fires. Similarly, the degraded areas located on the northern slopes of Sumaré, which are generally steep, increase the likelihood of fires in the area.

Thus, of all the areas, Sumaré (northern slope) possibly has a set of dynamic variables that positively influence fires - wind intensity and direction; temperature; relative humidity; humidity of easily combustible plant material (living and dead) (CARAPIÁ, 2006).

According to data from the Rio de Janeiro Municipal Works Department (2000), balloons were responsible for 11 of the 15 fires. Studies by Matos (2002) also show that balloons also caused fires, with around 23 outbreaks. However, of the 781 fires analysed in the massif, around 88% did not have the cause determined. According to these authors, other frequent agents are arsonists, the burning of rubbish and religious offerings. In only five cases, all in 1999, was the cause natural (lightning). Therefore, following the pattern observed in Brazilian conservation units (RAMOS, 1998), fire in the Tijuca National Park is caused by human beings and not by natural factors.

The release of balloons is a traditional habit among the people of Rio and takes place at any time of the year. However, they are more abundant during the June festivities, which coincide with Rio's dry season, when there is a significant increase between the afternoon and evening periods. (Secretaria de Obras do Munícipio do Rio de Janeiro, Available at: <http://www.dac.gov.br/imprensa/img40699.htm). More than fifteen years after the enactment of Law No. 9.605/98 (Environmental Crimes Law), article 42 of which makes it a criminal offence to manufacture, sell, transport or release balloons, some groups still persist in this action.

Therefore, as the main agents causing fire outbreaks in the Tijuca National Park and its surroundings (balloons, arsonists, rubbish burning and religious practices) have the potential to occur erratically throughout the year, an additional strategy to the fire history is to determine the areas that have characteristics that facilitate the ignition and spread of fire.

Until the beginning of this century, the lack of resources (human, technical, financial and infrastructural) for preventing and fighting forest fires meant that improvisation combined with the dedication of staff was the norm in the PAs managed by IBAMA (MORAIS, 2004).

In Rio de Janeiro, the Tijuca National Park has been the first to organise the brigade selection process, despite the fact that the others have a strong time constraint. Every year, 30 students are trained in a free course, who undergo an interview and preliminary tests of an eliminatory and classificatory nature (PLANO DE MANEJO/ PARNA- Tijuca, 2008).

If it is in the Park Manager's interest, vacancies are opened up to civil servants and/or employees working for the Conservation Unit, provided they have passed the Physical Aptitude Test (TAF) and the number of vacancies is not exceeded. In recent years, there has been an alternation of employees (Environmental Analysts and Environmental Technicians) in the role of Park Fire Manager.

5.1.6 - Exotic fauna and flora

Invasive Alien Species are organisms that, when introduced outside their natural distribution area, threaten ecosystems, habitats or other species. They are considered the second biggest cause of species extinction on the planet, directly affecting biodiversity, the economy and human health (MMA, 2006).

When these species are introduced into places where they don't occur naturally, they often don't find competitors or predators - consequently, their occupation and multiplication is facilitated and they end up threatening the permanence of native species, especially in degraded environments.

Invasive alien species cause damage not only to the natural environment, but also to the economy and health, and can have social and cultural impacts. More than 120,000 exotic species of plants, animals and microorganisms have already been recorded in six countries: South Africa, Australia, Brazil, the United States, India and the United Kingdom. Considering the number of exotic species that have already been identified in these countries, it has been estimated that a total of approximately 480,000 exotic species have already been introduced to the Earth's various ecosystems (SMA-SP).

Because PARNA Tijuca is located in the middle of the city, i.e. it is a forest, it is subject to permanent interventions by introduced fauna and flora of various forms and varying degrees of impact, which are still little known in the environment. Some of the most common and easily observed examples are jackfruit trees (Artocarpus integrifolia), dracaenas (Dracaena *arbórea)* and star tamarins (Callithrix penicillata).

According to the PARNA-Tijuca Management Plan, the main causes of the introduction of exotic species are:

• Invasion by domestic animals, mainly cats and dogs from homes in the immediate vicinity of the park. As the boundaries are not fenced, access is free for animals roaming the surrounding streets and roads.

• Abandonment of native and exotic species from religious rituals or not, which indicates intentional abandonment. In general, this occurs when the pet grows or breeds. These include different species of aquatic fauna, such as fish (carp, tilapia), crustaceans (freshwater shrimp), reptiles (turtles), puppies and kittens. Pigeons, goats and chickens are commonly left as offerings in religious rituals.

• Use of horses in the Pretos Forros area (sector D). The existence of a private stables around the area favours this practice.

• Visitors to the park accompanied by dogs, especially in the Paineiras area. This type of impact has never been assessed, but it can be predicted that populations of native fauna are subject to intense predation and the entry of unknown pathogens, which could cause serious damage to the biodiversity of this area. Many problems can be solved with effective monitoring, environmental education work and the erection of fences in areas where domestic animals regularly enter.

As far as the flora is concerned, the jackfruit tree Artocarpus integrifólia was one of the species introduced at the time of reforestation in the 19th century and its adaptation has been so successful that in some places it has become dominant. In these places, there has been a sharp decline in the sprouting of native species, forming veritable "jackfruit forests". Management of this species has been incipient, but if the importance of its management for wildlife is taken into account, it needs to be stepped up.

With regard to Panicummaximum Jacq, the increase is generally associated with the dynamics of fire, especially during the dry season. After a fire, the grass regrows quickly while the native vegetation, generally made up of slow-growing shrub and wood species, has its regeneration hampered by this rapid recovery. In this dynamic, new areas of forest are affected and replaced by grass with each fire (PLANO DE MANEJO/ PARNA- Tijuca, 2008).

This serious problem also occurs in the aquatic habitats in the Park area, despite the prohibition by specific legislation. Fish introductions have intensified recently in the state of Rio de Janeiro.

This has made it easier to obtain juvenile specimens of commercially cultivated species, many of which are exotic (MENDONÇA et al, 2004).

Studies on the occurrence and distribution of exotic fish species were carried out at points in the Park and in the following peripheral areas: Açude da Solidão, Rio dos Macacos, Rio da Gávea Pequena, Rio Itanhangá, Lago Frei Leandro and its access stream, the stream and lake of Parque da Cidade, the lake of Alto da Boa Vista, the lake of Alto do PARNA Tijuca, Rio da Tijuca, Rio da Cachoeira, Lago das Fadas, Rio das Pedras, Rio Trapicheiro, Rio da Barra, Rio Rainha, and the Ciganos river and dam (PLANO DE MANEJO/ PARNA-Tijuca, 2008).

It's important to note that some of the interviewees, a minority group, didn't consider exotic fauna and flora to be an environmental impact; on the contrary, they claimed that in nature everything adapts and harmonises.

5.1.7 - Hunting

Hunting is still practised in and around the Tijuca National Park. On the one hand, there are no more reports of large hunts, using groups of dogs as in the past, but on the other hand, there is still hunting using traps and jiraus. It can be said that a cultural habit persists, and that there is occasional subsistence hunting. The areas with the most traces of hunters are in the Tijuca Forest and Serra da Carioca sectors. In the Forest Sector, hunting is observed especially in the area of the Rio dos Ciganos and Cocanha basins and in the Serra do Andaraí (PLANO DE MANEJO/ PARNA- Tijuca, 2008).

In the Serra da Carioca sector, in the Cabeça and Macacos river basins, in the Gávea Pequena and Silvestre forests, in the Trapicheiros valley and in the Serra do Sumaré. The hunting season is usually during the hottest periods of the year. The most coveted species is the paca (Agoutipaca). The most hunted is the opossum (Didelphismarsupialis). Another type of hunting, which is quite representative, is that of passerines for wildlife trafficking. This occurs throughout the Park, especially in the areas with the greatest interface with urban occupation, mainly on the northern slopes of the Tijuca massif and on the slopes facing the Jacarepaguá neighbourhood (PARNA-TIJUCA MANAGEMENT PLAN, 2008).

5.1.8 - Transmission aerials

It is important to emphasise another problem that is conflicting for the protection of PARNA-Tijuca. This is not an external impact, but an internal one. It consists of the problem of the occupation of communications companies located on Morro do Sumaré:

There are records that the first authorisations to occupy areas of the Tijuca National Park for such activities date back to the early 1970s, when the former IBDF authorised, on a precarious and free basis, the first constructions and installations of towers and transmission equipment on the summit of the Sumaré hill. According to documentation presented by some of the occupants, around ten authorisations were granted from that time onwards, including public and private entities (Almeida e Peixoto, 1997, p. 56).

According to the same study, communication projects are one of the interferences or interrelationships between the urban area and the Tijuca National Park. It is a preferential and strategic area for the installation of broadcasting and telecommunications antennas and other related activities.

Aware that it would be impossible to make the proposed removal of these installations viable, as it would affect not only the radio and television stations, but also the public security agencies, the park's administration is working on several fronts to solve this problem, such as: intensifying inspections; and registering all the people who occupy areas of the park, with a view to regularising their occupation (in the update of the park's Management Plan, this area is characterised as a conflict area).

The management of PARNA-Tijuca necessarily involves partnerships with public institutions, research institutions, non-governmental organisations, residents' associations, the private sector, etc. The establishment of partnerships is vital for achieving the management objectives of the Conservation Area, given its uniqueness and complexity, as it is totally inserted in a metropolis (PLANO DE MANEJO/ PARNA- Tijuca, 2008).

During the questionnaires, some interviewees pointed out that the aerials had an impact on the environment. One type of physical pollution is visual.

5.1.9 - Irregular occupation (favelas) and general population occupation

The transformations of the different landscapes that occur on the earth's surface must be understood as the result of the dynamic combination of the role of biotic, abiotic and anthropic factors that interact dialectically with each other, becoming a single, inseparable whole of continuous evolution (TURNER 1989).

Among the changes to the vegetation cover for urban use, the emergence and growth of spontaneous favela-type urban features stands out in the Tijuca massif. These characterise the end point of the process of strong urban pressure resulting from population densification in the city as a whole (FERNANDES and et al, 1998). However, favelas are not necessarily the most important elements in the picture of contrasts that characterises the city's urban fabric, but rather the most visible.

Within the boundaries of PARNA-Tijuca, there are 43 low-income settlements (favelas) surrounding the conservation unit (UC) (ISER, 2000).

A peculiar characteristic of favelas is that they are generally located in less privileged places in terms of the likelihood of erosion problems, such as steep slopes at the foot of rocky outcrops. For this reason, it is of fundamental importance to evaluate the occupation process and its consequences in relation to the different hydrological and erosive responses in the massif, which are not uniform in space and time, and which constitute indispensable elements for investigations of these characteristics (FERNANDES and et al, 1998).

This massif is characterised by being a physiographic unit located within the urban site of the city of Rio de Janeiro, which is one of the major markers, along with the sea, of the process of expansion of the city's occupation. The city's population expansion meant that part of the population began to occupy parts of this massif from the last century onwards. Already in this century, especially in the second half, there has been a great proliferation of poorer housing in the form of favelas and other occupations of the formal city surrounding this massif, resulting in the current situation of strong urban pressure (PLANO DE MANEJO/ PARNA- Tijuca, 2008).

Favelas are an important variable in analysing the occupation of the Tijuca massif, as they occupy 4.6% of the total area of the massif. Their distribution along the massif is not uniform, given the uneven process of its occupation (FERNANDES et al, 1998).

The role of favelas in the advancement of the urban fabric is significant in the occupation of the area, although it takes on different magnitudes along the massif. However, it's important to remember that in the neighbourhood of sectors A (Alto da Boa Vista), B (Jardim Botânico and Gávea) and C (São Conrado), middle and upper class buildings are advancing towards the massif.

Despite being present in only 4.6% of the area of this massif, the incidence of slum occurrences is very high. These are also responsible for landslides. The distribution of landslide occurrences in the sectors revealed a close relationship with slum occupation, where in the areas with the highest concentration of slum occupation the number of occurrences in slums represented more than half of the total occurrences (FERNANDES et al, 1998).

In this sense, it is clear that there is a high concentration of landslides in these areas, because even though they are present in small areas, there is a high percentage of occurrences. On the other hand, in absolute terms, the formal city is the scene of a greater number of landslides within the massif than the favelas located there, so it's not just the favelas that are responsible for the landslides that occur.

As Fernandes (1998) puts it: this type of occupation plays a major role in the evolution of erosion processes in the massif, because as the favelas are installed "in tow" of the formal urbanisation process, they end up occupying the land that is neglected by the latter and more favourable to landslides, such as in the valley bottoms and at the foot of rocky walls, which are places where subsurface water is recharged. But this type of occupation is not the only one responsible for these processes, as other occupations such as luxury condominiums, middle-class homes and others also advance towards the massif and bring with them, to a lesser degree, problems of this type.

As for the scenario of violence, according to Peixoto et al (2006):

[...] the problem of violence around the PNT has increased significantly over time. The Map of Violence in Brazil, drawn up by the United Nations Educational, Scientific and Cultural Organisation (Unesco, 2004), reports that the state of Rio de Janeiro has seen homicide rates rise from 50.9 in 2002 to 56.3 in 2004, per 100,000 inhabitants, an increase of 10.6% when comparing the periods. The city of Rio de Janeiro has a rate of 64.2 per 100,000 inhabitants, in fourth place among the country's metropolitan regions. [...] Faced with this complexity, the management of the Tijuca National Park faces constant challenges. Among them are building a harmonious relationship between forest and city and integrating biodiversity preservation and social inclusion strategies. Thus, these are key issues for the Tijuca National Park to fulfil its basic management objective: "the preservation of the natural ecosystem, enabling scientific research and the development of environmental education and interpretation activities, recreation in contact with nature and ecological tourism", according to the definition of the national park category contained in Law 9.985, of 18 July 2000 - SNUC.

Although the population growth rate currently remains relatively stable in the municipality, the impoverishment of the population, combined with the search for housing near or close to places with paid work, has accelerated the historical process of informal slum construction, which began in the early decades of the 20th century, both on the slopes of the massif and in the surrounding lowlands (FERNANDES et al., 1999).

In the regional context, an alternative for the protection of protected natural areas is the implementation of ecological corridors and mosaic management. Mosaics are a group of PAs of different categories or not, close together, juxtaposed or overlapping, and other public or private protected areas. They must be managed in an integrated and participatory way, taking into account their different conservation objectives (integral protection units or sustainable use units) (BRASIL, 2000).

The intention, in this case, would be to make the protection of biodiversity, social interests and economic development compatible in the regional context, in accordance with the objectives of current legislation. The mosaic of PAs related to PARNA-Tijuca should therefore serve as a set of partners to protect the biodiversity of the municipal PAs and the park itself.

In relation to the water potential of PARNA-Tijuca, 63 springs in the The Tijuca Massif area supplies a small part of the population that benefits directly from the area. However, this same population can simultaneously cause and face problems in the same places and through improper use (PARNA-TIJUCA MANAGEMENT PLAN, 2008).

One of the elements enabling cooperative dialogue between PARNA-Tijuca managers and residents of the surrounding communities is the protection of water resources, as it is recognisable as a service provided to these people (CAMPHORA, 2005). However, the relationship has been conflictual, for various reasons, since the population is still not properly informed about the central issues related to possibilities and restrictions in relation to the management category represented by the park, as is the case in other full protection protected areas.

It is worth pointing out that a study more focused on individualising the different types of use, especially in terms of identifying non-favela urban areas, is of fundamental importance in order to assess more precisely the degree of influence of favelas on the various environmental impacts they have on the Tijuca massif.

5.1.10 - Noise and parties

This type of impact was pointed out by residents of the surroundings and buffer zone of PARNA-Tijuca, in sector A.

In Alto da Boa Vista there are many mansions, some of which have been transformed into party houses. These have caused problems for residents, such as irregular parking, excessive noise and the disposal of solid waste (rubbish) in the area.

The noise can also be attributed to the number of vehicles travelling along Estrada Menezes Cortes in Alto da Boa Vista. At peak times, according to residents, there are even traffic jams. This is mainly a type of physical pollution, noise pollution.

5.1.11 - Religious offerings

Another irregularity identified is the practice of religious rituals with offerings such as candles, clay bowls, drink bottles, leftover food (chicken, meat, farofa) and live animals (chicken, pigeons, goats). These practices are common both in the areas surrounding the Park, such as Avenida Edson Passos - Curva do S and inside PARNA Tijuca - mainly at the waterfalls in the Tijuca Forest and Serra da Carioca (Cachoeira do Quebra) (PARNA- TIJUCA MANAGEMENT PLAN, 2008).

In addition to environmental degradation, with an emphasis on river pollution and compromising the integrity of the landscape, the presence of organic rubbish in various places serves as food for wildlife and interferes with the dynamics of their populations.

According to the Conservation Unit's Management Plan, there are visible behavioural changes in the groups of coatis (Nasuanasua) found in the Tijuca Forest. It is possible that they take advantage of food offerings delivered by visitors. The disturbances caused to these animals should be assessed through scientific research and publicised to visitors, tourist guides and local residents.

However, there are ways of weighing up these practices, as they are part of the national culture. Brazil is a country of many beliefs, so religion in Brazil is very diverse and characterised by syncretism. The Constitution provides for freedom of religion and church and state are officially separated, Brazil being a secular state. Brazilian legislation prohibits any kind of intolerance, and religious practice is generally free in the country. According to the 2005 International Religious Freedom Report, drawn up by the US State Department, the "generally friendly relationship between religions contributes to religious freedom". Brazil is a religiously diverse country, with a tendency towards mobility between religions and syncretism[28] .

An alternative way of reconciling environmental preservation and guaranteeing the constitutional right to religious freedom was the Sacred Space of the Curve of the S Project, created by the State Secretariat for the Environment in August 2012, the aim of which is to regulate an area in the middle of the Tijuca Forest (surrounding PARNA-Tijuca), dedicated to the ritual practices of Afro-Brazilian religions that preserve nature. (State Secretariat for the Environment - SEA/ Projects and Programmes, 2012).

The Sacred Space of Curva do S, in Alto da Boa Vista, which has already been implemented, will have infrastructure that includes a guardhouse and gate, signalling totems, religious waste collectors and a religious waste treatment centre, with a compost bin, landfill and area for separating materials for recycling (PARNA-TIJUCA MANAGEMENT PLAN, 2008).

---

[28]Syncretism is a fusion of doctrines from different origins, whether in the sphere of religious or philosophical beliefs

There will also be a nursery - named the Garden of Sacred Leaves - for the production of seedlings of Atlantic Forest species donated by religious terreiros, to be used later by practitioners in rituals in the middle of the forest (PLANO DE MANEJO-PARNA-TIJUCA, 2008).

The so-called Curva do S, on Furnas Road, cuts through the Tijuca Forest and was chosen to host the first sacred space because, in addition to easy access, the management of the Tijuca National Park (PARNA-Tijuca) had already developed the Environment and Sacred Space Project 14 years ago. At the time, the demands and conflicts involving the intense religious use of areas in this conservation unit were mapped.

Set up by the government, the site will offer guidance to visitors and educational workshops on sustainable religious practices that use elements that don't harm nature, such as biodegradable products. Common in these types of religious rituals, materials such as bowls (clay dishes), crockery, glasses and bottles break easily and can cause injuries to people and animals. However, there are alternative materials that can be used in favour of environmental preservation, such as gourds, coconut or bamboo bowls, flowers and banana leaves.

Aimed in particular at tackling religious intolerance and environmental degradation encouraged by the inappropriate use of nature, such as the use of candles near trees, which can cause fires, the project is one of the aspects of SEA's work aimed at conserving the parks, combined with safety and preserving the Atlantic Forest.

With the launch of the Curva do S Sacred Space Project, a technical workshop will be set up to draw up guidelines and rules for the religious public use of protected areas. This will be followed by a workshop to build a participatory management plan for the area, with representatives from both the government and the religions that will be using the space.

In support of the environmental management of the area, clean-up and reforestation efforts will be carried out, with the planting of saplings, cultural events linked to dates in the religious and environmental calendar and seminars on the religious public use of protected areas. The Curva do S Sacred Space Project is the result of a partnership between SEA, Tijuca National Park and Rio de Janeiro City Hall, as part of the shared management of the park.

In addition to the Curva do S, regions of the Baixada Fluminense have already been researched for the project, since there is the highest concentration of Afro-Brazilian religious practitioners in the state.

5.1.12 -Deforestation

The loss of forested areas is intrinsically related to the forms of land use and the mode of production established in converted and deforested areas.

The relationships between social and economic variables and the destruction or preservation of habitats are highly complex, and small changes in the role of any of the historical, institutional or geo-environmental influences can lead to completely different results. In order to understand the social and economic causes of deforestation and identify the indicators that anticipate such pressures, it is necessary to include in the analysis the heterogeneity inherent in the anthropogenic process of territorial occupation. This complex interaction of factors is analysed according to the socio-economic determinants of land use in each of the main Brazilian biomes (Fundação SOS Mata Atlântica, 2005).

The history of deforestation in the Tijuca massif is quite peculiar. Chapter I of this work describes how the deforestation that altered the Tijuca massif took place through the coffee cycle, which began in 1760, and the reforestation process in the 19th century, motivated by the water crisis that plagued the city (ABREU, 1992).

As such, the Tijuca Forest is a secondary forest in an advanced stage of regeneration. And during its regeneration, the population of the city of Rio de Janeiro is growing at the same time. This presents a very contradictory scenario: what was reforested has been deforested, mainly by occupying the slopes of the Tijuca massif, and this occupation brings with it serious environmental damage. And another factor that makes PARNA-Tijuca, its buffer zone, surroundings and adjacencies more vulnerable to environmental impacts is the fact that it is an urban forest.

Considering that deforestation is the process of destroying forests through the action of man, we can consider that most of the environmental impacts surveyed in this research cause deforestation in the Tijuca National Park, as well as in its buffer zone, its surroundings and adjacencies.

5.1.13 - Excess visitors

In addition to unbridled urban sprawl, the Tijuca National Park, its buffer zone, its surroundings and adjacent areas suffer significant impacts due to the large number of visitors.

It is estimated that more than 1.5 million people visit it every year, with a high level of rubbish generation, depredation of its facilities, as well as compromising its water resources and biodiversity (PLANO DE MANEJO/ PARNA- Tijuca, 2008).

There are also problems relating to public use, such as the circulation of visitors by road, which the previous management plan sought to solve with the proposal to close access to motor vehicles, except for service vehicles. But it wasn't until 2008, on 27 March, that a small change was made, following a tender for a company to transport visitors in vans along the route between the Paineiras car park and Corcovado. But this is only for access to Corcovado (PARNA-TIJUCA MANAGEMENT PLAN, 2008).

In easily accessible areas such as the Horto waterfall, for example, there are various forms of deterioration caused by intense visitation. There is no proper supervision and it is common to find visitors bringing pet dogs, something that is forbidden in a Conservation Unit (BRASIL, 2000).

5.1.14 - Roads and Edge Effect

The forests that now cover the Tijuca National Park are the result of historical processes of expropriation, reforestation, the re-introduction of fauna and also the process of secondary succession that has taken place from vegetated areas in stretches that are difficult to access and possibly also from reforested stretches (PARNA-TIJUCA MANAGEMENT PLAN, 2008).

The Park is currently suffering from strong urban pressure (OLIVEIRA et al.,

1995), since fires, deforestation and processes derived from these are among the main causes of local degradation. In this context, edge effects are becoming more relevant.

Results found in similar formations subject to edge effects point to a less developed physical structure of the vegetation as well as lower biodiversity (BIERREGAARD et al., 2001), but various factors control the magnitude and distance of edge effects (HARPER et al., 2005), and thus plant responses associated with the existence of forest edges are necessarily site-specific.

The roads that cut through PARNA Tijuca promote a sectorised compartmentalisation of the vegetation and threaten the environmental integrity of the Park. Avenida Edson Passos, located in Alto da Boa Vista, separates the Tijuca Forest sector from the Serra da Carioca sector; Estrada das Canoas and Gávea Pequena separates the Serra da Carioca sector from the Pedra Bonita and Pedra da Gávea sectors; and Avenida Menezes Côrtes (formerly Estrada Grajaú - Jacarepaguá) separates the Pretos Forros and Covanca sectors from the Florestas sector.

Some of these roads (EDSON PASSOS and MENEZES CÔRTES) are extremely busy and make it difficult for terrestrial fauna to travel between sectors, which could lead to the insularisation of populations, directly affecting the biodiversity of this ecosystem.

Another factor to highlight is the edge effect on vegetation, which is mainly caused by atmospheric pollution. This effect in the Tijuca National Park compromises the vegetation, resulting in the growth of competing species such as vines, lianas and lianas whose weight on the treetops causes them to fall and consequently contributes to the erosion of the unprotected soil (IBGE, 1992). Studies indicate that the edge effect is characterised by an increase in the penetration of light and wind into the forest (KAPOSETe et al., 1997; BIERREGAARDET e et al., 1992), causing an increase in air temperature and an increase in the vapour pressure deficit inside the forest, which can lead to changes in species composition (LAURANCEe et al., 2006).

PARNA-Tijuca has seen unplanned and unmonitored trails in the border areas bordering the park, the expansion of areas of irregular occupation, urban densification at lower altimetry levels, and the intensification of the flow of vehicles and people. The proliferation of unplanned and unmonitored trails in the areas bordering the park are issues that jeopardise the very difficult sustainability of this large urban forest park today. It should be emphasised that although the Park category allows interaction between visitors and nature and the development of recreational, educational and environmental interpretation activities (BRASIL, 2000), these activities must be controlled and follow the rules and restrictions established in the unit's management plan, with a view to managing natural resources and sustainable development.

5.2- ENVIRONMENTAL PERCEPTION OF THE RESIDENTS OF THE BUFFER ZONE, SURROUNDINGS AND ADJACENCIES TO THE TIJUCA NATIONAL PARK:

The last part of the questionnaire was answered only by residents of the buffer zone, surroundings and neighbourhoods of PARNA-Tijuca (Figure 18).

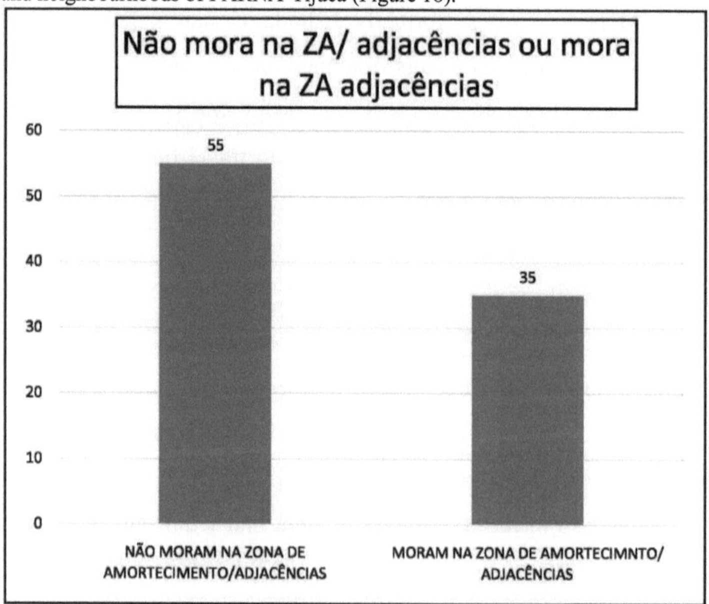

Figure 18 - does not live in ZA/ adjacencies - lives in ZA adjacencies

The data reported here comes from the final part of the questionnaire used in the research, which was to be answered exclusively by residents of the buffer zone, surroundings and adjacencies of PARNA-Tijuca. This data is highly relevant because the interviewees are witnesses to the transformations that have taken place over time, living on the edge of the forest and living in the contrast between the natural environment and urbanisation. We analysed their knowledge of the ZA, surroundings, adjacencies and core area of PARNA-Tijuca, the value they attach to living in the locality, how they identify environmental impacts and how they compromise their lives and local biodiversity. Below is a profile of some of the groups that occupy the slopes and areas below. This group totalled 35 interviewees. This corresponds to 39 per cent of the sample. In this group there is a variation in social class, level of education and history of occupation in the region. Most of them live outside Sector A, but even to a lesser extent there are residents outside Sectors B, C and D.

In sector A, most of the interviewees live in communities in Alto da Boa Vista. Two communities will be highlighted because field research was carried out in these localities. These communities are Vale Encantado and Furna de Agassiz. Even though some interviewees live some distance from PARNA-Tijuca, such as Rua Conde de Bonfim, Rua Barão de Itapagipe and Rua Mariz e Barros, all in Tijuca, they answered the questionnaire considering themselves to be residents of the ZA. We can see that the interviewees generally don't understand the boundaries of PARNA-Tijuca.

According to Simões (2007): Knowledge of the totality of a municipality, therefore of the totality of its places, is the privilege of a few, of those endowed with such mobility as to enable them to experience the different places from their own place. Even in very small municipalities, with very few exceptions, this premise is true.

5.2.1 - The Enchanted Valley Community

The Vale Encantando community is located in the PARNA-Tijuca buffer zone. It is home to a population of approximately one hundred and twenty residents (Cooperativa Vale Encantado - COOVE). The characteristics

arc quite peculiar, it seems that there is a harmonious interaction between this population and the forest. Most of the houses are far apart, so they don't form agglomerations.

At the same time as PARNA Tijuca was inaugurated in the 1960s, flower growing was the main alternative for generating income for the local population and achieved great development, but the activity died out at the end of the 1970s due to competition with the Dutch and German colonies established in the mountains of Friburgo, Teresópolis and Petrópolis. Prior to this activity there were coffee plantations, and the community pre-dates the founding of the park (VALE DO ENCANTADO. 2011).

It was during this period that a new economic activity began in the region with the installation of black granite quarries. This activity attracted a large number of new residents due to the increased supply of labour and jobs. However, the extraction of granite caused serious damage to the environment due to explosions and the activity itself, leaving the land without vegetation cover and affecting the many springs that exist in the area. Due to growing concern about environmental quality and dwindling granite reserves, production slowed down and stopped for good in 1990. For this reason, many residents left the valley in search of better employment opportunities and infrastructure.

In recent decades, due to the housing shortage, several favelas have been built in the vicinity of the Tijuca National Park, challenging public authorities and local residents to guarantee the protection and conservation of the forest.

In a joint effort between public authorities and communities, through the Alto da Boa Vista Citizens' Council (CONCA), attempts are being made to recover the environment from the consequences caused by agricultural and granite exploitation in the area. The community is looking for sustainable measures to generate income and preserve the environment.

Eco-tourism is a successful activity in the Enchanted Valley, which is part of the city's tourist route. In 2005, the French solidarity organisation ABAQUAR/PARIS proposed a partnership to the Vale Encantado community to support the structuring of a social and environmental cooperative. This support gave rise to the Vale Encantado

Cooperative (COOVE), which aims to strengthen the productive and entrepreneurial potential of the community through sustainable tourism initiatives and the development of local gastronomy.

Today, the Vale Encantado community has a team of 20 co-operators who act as guides on the ecological trails and run the restaurant and social buffet. The article below, from O Globo newspaper's Razão Social magazine, illustrates the importance and dynamics of the Vale Encantado community (Figure 19):

The stories told about the more than 40 favelas that grew up around the Tifuca Forest in Rio de Janeiro speak of the violence of drug trafficking or the environmental damage caused by its expansion. But in the community of Vale Encantado, in Alto da Boa Vista, residents are writing a different story, one that includes environmental preservation and income generation through tourism, cookery and arts activities. A cooperative they set up seeks out companies as partners and mobilises the community for other achievements, such as better health care.

Figure 19 - article in the magazine Razão Social, in the newspaper "O Globo", 15 September 2009.

The community, however, is under threat of expropriation by the government because it lies within the buffer zone of the Tijuca National Park. Most of the residents were born there.

The town was originally closely linked to a granite quarry. With the end of this activity, the local population

decreased. The people who remain have witnessed the transformation of the environment from degraded to recovered. In general, this population's view of environmental preservation is very positive.

The main environmental impacts they observe are: old trees, dry trunks, burning, hunting and erosion, excessive solid waste (rubbish), the presence of a club, a residential condominium (visual pollution, contrasts with the forest landscape, as well as generating the environmental impacts typical of any human occupation), deforestation and visitors from outside. As for the condition of the rivers, they generally consider most of them to be very polluted.

The positive aspects of the locality pointed out by the residents are related to the tranquillity, fresh air, silence that contact with nature provides and the issue of security, as so far there has never been any organised crime or any similar threat. The negative aspects relate to the lack of transport, access to shops and basic sanitation. The only form of public transport up to the community is provided by a resident in a passenger car.

The actions that the residents suggest and their expectations for local environmental preservation and recovery are: for human beings to have more love for the environment, clean-up efforts, more awareness, environmental education projects, monitoring, incentives to set up native seedling nurseries, reconciling income generation and environmental preservation, information for the population and setting up a biodigester.

Figure 20 - houses in the middle of the forest.

5.2.   2- The community of Furnas de Agassiz:

The community of Furnas de Agassiz lies between sectors A and C, outside the Buffer Zone, but on the edge of it. Access is via the Furnas road, near number 3001.

This location is home to unique natural resources, but they are in a state of disrepair because irregular occupation has taken place on the waterfall's bed. There is a waterfall in the area, but it looks polluted, as evidenced by its visual appearance and unpleasant odour. In addition to so many other things, there is a beautiful speleological heritage. From the evidence of the buildings that remain, it seems that the site was revitalised in 1905 by the Pereira Passos government, which can be seen in the dated buildings that are still there, even if they are damaged. From the state of deterioration, it can be concluded that there has been no intervention by the public authorities to ensure the conservation of the area. There is also a ruined senzala without any identification (Figure 21).

Figure 21 - ruined slave quarters.

In order to maintain the environmental sustainability of the areas bordering PARNA-Tijuca, Decree No. 11.301 (of 21 August 1992) created the APARU (Environmental Protection and Urban Recovery Area) of Alto da Boa Vista, covering most of the basins of the Maracanã and Cachoeira rivers (GEOHECO, 2000).

This Conservation Unit was only regulated in April 2003, by Bill 1.307, with the aim of re-establishing connectivity between the forest fragments and the contiguous and continuous forest area of the Tijuca National Park. The aim was to guarantee the system's functionality by preserving water sources, regulating the microclimate and maintaining the stability of the slopes (COELHO NETTO, 2009).

In this location there are some very curious geological formations in the Tijuca Forest, the Furnas de Agassiz, or simply Furnas. Formed by rocks formed in the Palaeozoic period, the set of caves made up of large stone slabs, interspersed with lakes and waterfalls have enchanted visitors since the time of the arrival of the royal family when they were discovered by members of the artistic mission, including Taunay, who settled in a relatively nearby area (LEAL, 2003).

In the second reign, it was a stopping point for adventurers and explorers who travelled through the area where the park would be created in the future, and in the 20th century, with the improvement of access roads, it became one of the most popular spots for country picnics, becoming a tourist attraction (LEAL, 2003).

In the 1960s and 1970s, it became nationally known for having served as a backdrop for countless film and television productions, such as the famous Cuca's lair in the first version of Rede Globo's Sítio do Pica-Pau Amarelo and also as the King Solomon's Mines in Trapalhões. (LEAL, 2003)

The area seems to be in a state of great vulnerability. As you walk around the area, you can see the religious offerings, left in a careless manner by their devotees. And what seems to be the biggest threat is the rapid process of favelisation that has accompanied the slopes of the Furnas Road facing Barra.

The Mata Machado and Fazenda shantytowns dump a large amount of sewage, which, together with the remains of offerings, transforms the once pleasant area into a polluted environment. The Cachoeira River arrives at the site totally polluted and with its flow greatly reduced, the Grande Waterfall, opposite the famous Cascatinha, is today with its flow reduced and its waters polluted, which creates a shock when we see the pit in the rock still bare and the paintings made by the artistic mission and photos from the beginning of the 20th century.

The residents interviewed have lived in the community for many years or were born there. They consider it a good place to live. They point out the following positive aspects: water collection from the spring (clandestine), there is no drug trafficking, contact with nature and the place is quiet, there is no noise pollution. However, the negative aspects include intense environmental pollution caused by sewage being dumped in the river and excessive rubbish. This attracts vectors of various diseases, such as dengue fever, leptospirosis and others. Rubbish collection is very precarious in the locality, there is no regular rubbish collection by COMLURB, causing rubbish to accumulate. They also report that over time various species of fauna (mainly fish) and flora have become extinct.

The interviewees reported many changes in the area, such as the reduction of the green area and the increase in housing. No official data was found on the number of residents, but according to the interviews there are around 300 houses. Most of these are high-risk constructions that border the riverbed.

There are few studies on the area. The situation is very critical and there is an urgent need for civil society and public authorities to take a responsible approach. There is a great need to enforce laws that condemn the various visible irregularities and environmental crimes. This area is regulated by the government as a specially protected area, it is included in an APARU, a category of sustainable use conservation unit, but there is no supervision, no proper management, no preservation initiative. Residents claim that there are no improvements in the area and believe that basic sanitation, cleaning up the rivers and improving solid waste collection are necessary.

5.2.3 - Other Communities Alto da Boa Vista

The low-class communities that form in the Massif are similar in terms of their irregular occupation, but each one has its own history of formation. There are few studies on the peculiarities of these communities. Other residents of Alto da Boa Vista were interviewed and the fact that they consider the area a good place to live is practically unanimous.

Residents of the Tijuaçú community, Estrada do Soberbo, Caminho Rodrigo Silva, Estrada do Açude and Rua da Cascata were also interviewed. They claim that the population lacks transport, there is an absence of public authorities and rubbish collection is precarious. The main impacts they see on the environment are: exotic species, cars, car parks on the sides of the roads, the presence of houses in the river protection area, the accumulation of rubbish in the rivers and in many places.

5.2.3.1 - Tijuca residents:

When answering the questionnaire, some people who live in Tijuca considered themselves to be residents of the area outside PARNA-Tijuca. As the survey extends to areas beyond the Buffer Zone, these answers were validated because they consider it to be very restricted. This group belongs to the middle class and chose their place of residence according to the advantages it offered them. And they have a relationship with PARNA-Tijuca, either professionally or for leisure purposes.

The sample included residents of Conde de Bonfim, Barão de Itapagipe, General Rocca and Visconde de Cairú streets. The interviewees considered the areas where they lived to be very favourable in terms of urban transport, services in general, shops and proximity to their workplace. As interviewee César Augusto Ferreira, 52, put it: "It's a central area, with lots of shopping and transport options. Pleasant climate (due to the proximity of the Forest), easy access (transport) and plenty of entertainment."

The negative aspects raised were a lot of noise, violence, heavy traffic, poorly maintained pavements and streets, lack of security, close proximity to communities, no selective waste collection, lack of activities in the surrounding communities, little preserved green space, a lot of pollution caused by means of transport.

The main environmental impacts they point to are: noise pollution, rubbish, air and water pollution and the felling of trees, a change in the micro-climate, heat (due to the removal of trees and tarmac, leading to poor drainage). They consider the rivers to be in a very poor state of conservation.

The actions they believe are necessary are: commitment from public authorities, treatment of sanitary sewage, effective monitoring of sewage and rubbish dumped illegally, improvement of surface water drainage, afforestation, landscaping of streets and squares. Cleaning up the rivers, removing the shackles that pour water into manholes (rainwater) and rivers (in many places sewage is dumped "in natura"). Residents identify the need for people to be more aware and for the area covered by vegetation to be expanded.

5.2.3.2 - Sector B:

The sample from sector B is quite small and the interviewees live in privileged areas of Laranjeiras and Cosme Velho. They consider it a good place to live because it's quiet, there's lots of vegetation and wild animals. It also has a pleasant climate and tranquillity. The main problems they point out are: visual pollution, vehicles, lots of cars parked on the streets. This pollutes the air and causes urban disturbances.

The main change pointed out over time, which fits in with an environmental impact, is slum clearance on the hillsides. The suggested action for improvement is to remove the slums and reforest them. Putting up walls or fences to protect the forest.

5.2.3.3 - Sector C:

The community interviewed in the area outside PARNA-Tijuca was Vila Canoas (Figure 33), on Estrada das Canoas, in São Conrado. The Vila Canoas community begins on the edge of Estrada das Canoas (Figure 36). According to the interview data, the occupation was initially irregular, but the city council regularised all the housing.

According to the data provided by the interviewees, the community has undergone many improvements over time. Today there is tranquillity, security, water, electricity, sewage treatment and good neighbourliness. Unlike in the past when there was no health centre on site, nor even a water and electricity supply, the streets were not paved, and there were many shacks. They claim that what's missing is a recreational area for the children. They don't identify any environmental impacts on the site, and suggest that there should be more social projects aimed at environmental education.

We can deduce that these improvements can be attributed to the fact that Comunidade canoas is located in the South Zone, on a road (Estrada das Canoas) where there are many mansions and luxury condominiums.

5.2.3.4 -Sector D: Jacarepaguá

As stated at the beginning of this chapter, there was no field research, neither in the core area nor in the external area of this sector. The interviewees living in the external area were interviewed via the internet and in the core area of sector A. They are residents of Estrada de Pau Ferro and Estrada Buganville.

Residents say that there is a lot of green area, with forest fragments, where you can enjoy fresh air, rivers, vegetation, cleanliness and clean air. However, the environment is compromised by the amount of solid waste generated, urban disorder and growing irregular population occupation, pollution from vehicles, burning, hunting, clandestine water abstraction and the introduction of exotic species. River pollution is precarious, and sewage is dumped directly into the springs.

Violence in the communities along the Grajaú-Jacarepaguá road is also pointed out as a factor that compromises the quality of life around PARNA-Tijuca in sector D.

The changes observed over time reflect the population growth to which the city of Rio de Janeiro is subjected. Older residents say that the place used to be quieter, since the population was smaller.

They suggest the following actions for local improvements: river and lake drainage and maintenance, implementation of basic sanitation, selective collection and environmental education projects.

CHAPTER 6

CONCLUSIONS

The Tijuca Urban Forest (PARNA-Tijuca) is a highly vulnerable protected natural area. It is a complex administrative unit to manage given the challenges it faces, the main one being the continuous anthropic pressure (action by people from different social groups) on its boundaries. There is also violence and social exclusion in the surrounding areas, especially in the favelas. This is why it is essential to develop initiatives and strategies aimed at building citizenship and social inclusion, consolidating a new way of acting aimed at protecting renewable resources, fauna and flora. In this case, the interaction with the surrounding area is very different compared to protected areas located far from large metropolises, i.e. in inland areas. Management must therefore be differentiated, seeking more efficient measures in terms of urban occupation, surveillance, etc.

The environmental impacts pointed out in the surrounding buffer zone in this study represent serious threats to the conservation of the Tijuca Forest, and studies to better characterise and minimise them are of the utmost importance. As well as a detailed diagnosis of the area to be defined as a Buffer Zone, so that its definitive limits are consistent with what the legislation determines, in terms of controlling and monitoring actions that potentially impact natural resources within the legally protected area. There is a preventative nature to the buffer zone. If it serves as a filter against external attacks on the conservation unit, then it serves to prevent any kind of degradation that could jeopardise the integrity of the core area.

In order to guarantee the preservation of the core area of PARNA-Tijuca, which seems to be better preserved, preventive and maintenance measures are needed in the surroundings and buffer zone. An example of a common impact are the roads inside PARNA-Tijuca, where a large number of private vehicles circulate. There is also a lack of basic sanitation and poor solid waste management.

It can be seen that the buffer zone proposed by the PARNA-Tijuca Management

Plan is very restricted, as it is located very close to the boundary of the Conservation Unit. There is a great diversity of fauna, flora, speleological riches, historical monuments and water sources outside the boundaries of the buffer zone, which are already highly degraded.

Research is needed on the communities surrounding PARNA-Tijuca. According to what was observed in the communities researched in this paper, it can be seen that they have largely lost their memory and their history of formation. Each of them has its own peculiarities in terms of formation, history and general characteristics. They have their own identities, but there are no official records of them. Information is passed down from generation to generation. And they are considered marginal due to the way they are occupied and the lack of resources (transport, sanitation, etc.) by their self-concept and by society.

One of the reasons why the buffer zone is so restricted is the number of communities around PARNA-Tijuca. It is important that there are more effective forms of integration between the park and the surrounding population. The development of strategies for social integration and improving the quality of life of the surrounding low-income populations.

Studies of these communities, legalisation of housing, occupation control, alternative housing in the case of risky occupation, sanitation and revitalisation policies, improved transport, partnerships with NGOs and educational institutions for environmental education projects and income generation with local resources on the part of the state are all important measures. This will lead to the appropriation and valorisation of the space by this population. This will reduce the environmental impacts that threaten the core area of PARNA-Tijuca by raising awareness. Some of these communities are part of APARU - Alto da Boa Vista (Annex I), but this specially protected area is neglected and the guidelines are not enforced. The government created the APARU- Alto da Boa Vista, but its actions are not effective. The threats to biodiversity are still intense and ongoing.

The majority of those interviewed who live in middle and upper class housing think that favelas are an environmental impact and that the immediate measure (two people in the sample had this opinion) is to remove all the favelas from the slopes of the Tijuca massif. This is the most visible problem. But it's not the only one. Improvement measures are needed, but with a balance, efficient public policies and compliance with the law.

There are very important actions, such as volunteer work at PARNA-Tijuca. It's a way of improving the management and administration of the conservation unit through meaningful activities, generating benefits for both the park and the volunteers, in an opportunity for interaction and reconciliation between man and nature. The objectives are also: to develop environmental education through voluntary actions aimed at raising awareness and conserving the park; to encourage and develop positive territoriality between the visitor and the conservation unit through volunteering; to consolidate the participation of society through volunteer work in conservation units; to create opportunities for the exercise of citizenship through volunteering in PARNA-Tijuca; to promote participatory awareness among volunteers, visitors and residents of the park's surroundings. (AMADOR, 2013).

Common sense also makes it difficult to identify the spatial limits of the park. The community in the region is unaware of the importance and function of a National Park and is distanced from it on a daily basis. It is therefore necessary to develop ways of improving this relationship between society and PARNA-Tijuca.

In short, PARNA Tijuca is located in the middle of the urban fabric of the city of Rio de Janeiro. It is recognised as one of the largest urban forests in the world. This makes it necessary to adopt a management model that integrates urban, environmental, social, economic and cultural factors. To this end, it is necessary to use various instruments in a coordinated manner, with the local sphere of action emphasising the rules relating to land use and occupation. The surrounding areas will therefore have to be limited in order to organise, guide and promote compatible activities, taking care not to make the neighbouring communities economically and socially unviable. At the same time, these actions should protect PARNA Tijuca from "edge effects", which cause degradation from the edge of the forest towards its interior (GEOHECO, 2003).

The identity of the city of Rio de Janeiro is directly linked to PARNA-Tijuca.

Part of the population is unaware, for example, that Corcovado is part of the park and that the Tijuca Forest is its administrative unit, therefore subject to the specific rules of the objectives of a national park. Many landscape references and postcard scenes from Rio can be found in the Tijuca massif. Corcovado, Christ, Pedra da Gávea, Vista Chinesa, Cascatinha, the Forest, Paineiras, Pico da Tijuca, Mirante Santa Marta, all of which combine great visitation with proximity to the population's identity.

This research also fulfils one of the SNUC's guidelines, article 5, item IV: The SNUC will be governed by guidelines that: "seek the support and cooperation of non-governmental organisations, private organisations and individuals for the development of studies, scientific research, environmental education practices, leisure and ecological tourism activities, monitoring, maintenance and other management activities of conservation units"

And yet there are great contrasts, even in the face of so much beauty and natural wealth. It's a scenario of paradoxes that needs the attention and action of all social sectors.

Below are some excerpts from the song, Rio 40 Graus, which portrays the contrasts of the city of Rio de Janeiro and fits in well with the reality of the Tijuca Massif, which lies between beauty and chaos.

"Rio 40 degrees
Marvellous city
Purgatory of beauty
And the chaos...
Hot-blooded capital
From Brazil
Hot-blooded capital
The best and the worst
Brazil...
Hot-blooded city
Mutant wonder...
Rio is a city
Of mixed cities
Rio is a city
Of camouflaged cities
With mixed governments
Camouflaged, parallel
Sneaky
Hiding commands...
Who owns this bank?
Who owns this street?
Whose building is it?
Whose place is this?

This place is mine
This place is ours
I want my badge
I'm from Rio de Janeiro, Brazil.
In the hot-blooded city
In the mutant city of wonder..."

(Rio 40 graus- Fernanda Abreu/ Fausto Fawcett) - Abreu on the 1992 *album* Sla2 Be Sample).

CHAPTER 7

BIBLIOGRAPHY

ABREU, Maurício Almeida. **The city, the mountain and the forest**. In: ABREU, M.A. (editor) Natureza e Sociedade do Rio de Janeiro, Biblioteca do Rio de Janeiro, Secretaria Municipal de Cultura, Turismo e Esportes. p. 54-103, 1992.

___________ . **Reconstituting a forgotten history**: the origin and initial expansion of Rio de Janeiro's favelas. In: Espaço e Debates, 14(37), p. 34-46, 1994.

EvRioçde urRana do RíR de JaReiro. Rio de Janeiro: IPLANRIO/ Zahar, 1987.

ALENCAR, José de. *ORrod completas*. Rio de Janeiro: Aguilar, 1959.

ALIE, Joan Martínez. **The Ecologism of the poor**: environmental conflicts and languages of valuation. São Paulo: Contexto. 379 p., 2007.
AMADOR, André Bittencourt; PALMA, Lúcio Meireles. **Ten years of the Tijuca National Park volunteer programme**. In: Anais, Uso Público em Unidades de Conservação. N, 1. V,1. 2013.

O ECO ASSOCIATION. **Portal o eco**. Accessed on: Apr. 2013. Available at: <http://www.oeco.org.br/index.php?option=com>. 2011

AZEVEDO, Cristina. **Invasive exotic species**. In: State Secretariat for the Environment. Biodiversity and Natural Resources Coordination Office. Riparian Forest Notebooks Riparian Forest Recovery Project Coordination Unit. Accessed on: 01 April. Available at: <http://www.ambiente.sp.gov.br/>. São Paulo: SMA, 2009.

BANDEIRA,a, C. M. **Tijuca National Park**. São Paulo: Makron Books, 1993.

BARROS, Gabriel. **Resident of the Vale Encantado Community**. Interview with the author, 5 September 2013.

BELINATO, T. A. &SILVA MATOS, D. M. **O Impacto de Pteridiumaquilinumvar.arachnoideum, Pteridophyta, na Germinação e Morfologia de Plântulas de Espécies Arbóreas da Mata Atlântica**. In: Anais do VI Congresso de Ecologia do Brasil, Fortaleza, pp. 381-382, 2003.

BENCHIMOL, Jaime L. *Pereira Passos: a tropical Haussmann*. Municipal Department of Culture. Rio de Janeiro, 1990.

BIERREGAARD, R.O.Jr; GASCON, C.; LOVEJOY, T.E.; MESQUITA,R. (eds.). **Lessons from Amazonia**: The Ecology and Conservation. 2001.

BIERREGARD, R. O.;Lovejoy, T. E.; Kapos, U.; Santos, A. A.; Hutchings, R.W. **The**

biological dynamics of tropical rainforest fragments. Bioscience. 1992

BRAZIL. Ministry of the Environment. *Federal Law No. 9.985/2000 - National System of Conservation Units*. Brasília, 2000.

___. **law 12.305, of 2 August 2010**. National Solid Waste Policy.

___. **Report of the Ministry of the Empire/1844**.

___. **Report of the Ministry of the Empire/1874**.

. **Law No. 9.605, of 12 February 1998**. Presidency of the Republic - Civil House.

Decree no. 60.183.

Law No. 12.651, of 25 May 2012. Presidency of the Republic

. PLC - House Bill No. 30 of 2011. Senate.

CHAMBER OF DEPUTIES.**PL 1876/1999**.

CARAPIÁ, V. R. **Prediction of the Fire Risk Index and Computer Modelling of the Behaviour of the Advancing Fire Front in the Tijuca Forest National Park.** PhD Thesis, Engineering Postgraduate Programme, UFRJ, 2006.

COELHO NETTO, A.L. Hidrologia de Encostas na Interface com a Geomorfologia. In: **Geomorfologia: Uma Atualização de Bases e Conceitos**, organised by GUERRA, A.J.T. e CUNHA,S.B.; Ed. Bertrand Brasil, chap.3, p. 93-148, 1994.

_______________ . **The Geoecosystem of the Tijuca Forest**. In:*Abreu; M.A. de (Org): Nature and Society in Rio de Janeiro*.RiodeJaneiro: Biblioteca Carioca. Municipal Department of Culture, Tourism and Sport. Chap.5, p. 104-142, 1992.

CONAMAN. **Conama Resolutions No. 357/2005**, Conama No. 274, Conama No. 344/2004, and Ordinance No. 518 of the Ministry of Health.

COARACY, V. **Memórias da Cidade do Rio de Janeiro**. 2. ed.Rio de Janeiro: José

Olympio, 1965.

___. **Rio de Janeiro in the 17th century**. Rio de Janeiro: José Olympio, 1944.

CONAMA. **CONAMA Resolution No 001/1986.** 1986

___. **CONAMA Resolution No 13/1990.** 1990

Stockholm Declaration - 1972 - mpba.mp.b

DEAN, Warren. *A -erm - -ogo - A* história e a devastação da Mata Atlântica brasileira.2ª ed- São Paulo:Ed. Companhia das Letras, 1998.

DIAS, Pessoa. **Analysis of the interconnection of the sanitary and rainwater sewage systems in the city of Rio de Janeiro:** Valuing water collections from a systemic perspective.Dissertation (Master's Degree in Sanitary and Environmental Engineering). State University of Rio de Janeiro-Centre of Technology and Sciences Faculty of Engineering Department of Sanitary and Environmental Engineering, 2003.

DRUMMOND, José Augusto. **The garden inside the machine**. Estudos históricos, Rio de Janeiro, v.1,n.2, 1997.

____. **Devastation and environmental preservation in Rio de Janeiro**. EDUFF. Niterói, 1998.

Decree No. 11.301 (of 21 August 1992). Creation of APARU-RJ.

Integrating environmental protection and social participation in urban areas. Socio-environmental diagnosis of the Tijuca National Park and surrounding areas.ENVIRONMENTAL EDUCATION. IN CONSERVATION UNITS. BRAZILIAN INSTITUTE OF SOCIAL AND ECONOMIC ANALYSES (IBASE). ACCESSED ON: APR. 2013. AVAILABLE AT <WWW.IBASE.BR>. 2006

____. **PARTICIPATORY MANAGEMENT IN PROTECTED AREAS.** IN: ENVIRONMENTAL EDUCATION. IN CONSERVATION UNITS. BRAZILIAN INSTITUTE OF SOCIAL AND ECONOMIC ANALYSES (IBASE). APR. 2013. AVAILABLE AT <WWW.IBASE.BR>. 2006

BRAZILIAN INSTITUTE OF SOCIAL AND ECONOMIC ANALYSES (IBASE). **Socio-environmental diagnosis of the Tijuca National Park and surrounding areas.** AVAILABLE AT: <WWW.IBASE.ORG.BR>. ACCESSED ON: APR 2013. 2006

____. **TIJUCA NATIONAL PARK. AVAILABLE** AT: <WWW.IBASE.ORG.BR>. ACCESSED ON: APR 2013. 2006

. WATER - Public goods in conservation units.

FEEMA, 2004. **Annual Air Quality Report 2003**. Environmental Planning Department, Air Quality Division, Rio de Janeiro.

FERNANDES, N. F.; GUIMARÃES, R. F.; GOMES, R. A. T.; VIEIRA, B. C.; MONTOGOMERY, D. R.; GREENGERG, H. **Topographic Control of Landslides** in Rio de Janeiro: Field Evidence and Modelling. CATENA, Amsterdam, vol. 55, pp. 163-181. 2004.

FERNANDES, M.C., LAGUENS, J.V.M. and COELHO NETTO, A.L. (1999) - **The process of occupation by favelas and its relationship with landslide events in the Tijuca massif.** UFRJ Geosciences Yearbook.

FERNANDES, M.C. **Geoecologia do Maciço da Tijuca:** Uma abordagem geo-hidroecológica. Master's dissertation, Postgraduate Programme in Geography, 1998.

FERNANDES, M. C,; LAGUÉNS, J.V. M.; NETTO, A. L. C.**The process of** occupation by favelas and its relationship with landslide events in the Tijuca Massif/RJ.

Queiroz Fernando Fonseca de. **Brasil: Estado laico e a inconstitucionalidade da existência de símbolos religiosos em prédios públicos.**JusNavigandi. 2005

FERRARA, L. *Olhar* **periférico: linguagem, percepção ambiental.**2 ed. São Paulo: Editora da USP, 1999.

FOLHA DE SÂO PAULO. **Government and ruralists differ on code.** Folha de S. Paulo, São Paulo, 25 Apr. 2012.

GALINDO-LEAL, Carlos. SOS Mata Atlântica Foundation. **Mata Atlântica**: biodiversidade, ameaças e perspectivas / edited by, Ibsen de Gusmão Câmara; translated by Edma Reis Lamas. - São Paulo: SOS Mata Atlântica Foundation - Belo Horizonte: Conservation International, 2005.

GREY, W.G; DENEKE, F. J. JoJm Wiley & Sons; YDBERG, D.; FALCK, J. **Urban Forestry in Sweden from a silvicultural perspective**: a review. LandscapeandUrban Planning.V.47 n.1-2, 2000. p.1-18., 1986.

GUERRA, A.J.T.; CUNHA, S.B. da (Org.) **Impactos Ambientais Urbanos no Brasil.**Rio de Janeiro: Bertrand Brasil, 2001.

HELMBOLD, R.; VALENÇA, J. G; LEONARDOS, Jr, O. H. *Geological Map of the State of Guanabara,* scale 1:50.000. Rio de Janeiro, DNPM/MME, 1965.

HARPER, K. A.; MACDONALD, S. F.; BURTON, P.J.; CHEN, J.; BROSOFSKE, K. D.; SANDERS, S. C.; EUSKIRCHEN, E. S.; ROBERTS, D.

& EESSEN, P. **A.Edge influence on forest structureand composition in fragmented landscapes.**ConservationBiology, **19(3)**: 768-782., 2005.

HEYNEMANN, Cláudia. **Tijuca Forest** - Nature and Civilisation. Municipal Department of Culture. Rio de Janeiro, 1995.

HYDE PARK .SENIOR PLAYGROUND FEASIBILITY STUDY, 2012

HOLANDA, AURÉLIO BUARQUE F. **Dicionário da Língua Portuguesa**. 5ª ed. Editora Nova Fronteira, 2000.

IBASE. *Socio-environmental Diagnosis of the NLcional dL T^uca Park. Action Line 2.1 - Restructuring the Park's Advisory Council.* Petrobras Environmental Programme - Water in Conservation Units Pilot Project for the Atlantic Rainforest: Terrazul Institute - IBASE - Tijuca National Park, Rio de Janeiro, 2005

IBGE.1993, **Vegetation Map of Brazil.** Brazilian Institute of Geography and Statistics. Rio de Janeiro. 1993.

___. 2007. **Population Count 2007**. Brazilian Institute of Geography and Statistics -

Ministry of Planning, Budget and Management. Available at http://censos2007.ibge.gov.br/, accessed on 20/02/2008.

___. **Technical manual of Brazilian vegetation**. Brazilian Institute of Geography and Statistics, Rio de Janeiro. p. 18. 1992

IEF. 1994. **Atlantic Forest Biosphere Reserve in the State of Rio de Janeiro**. Zoning Map. State Secretariat for the Environment and Special Projects - SEMAN. State Forestry Institute Foundation (IEF). Government of the State of Rio de Janeiro. Rio de Janeiro.

ISER-Institute for the Study of Religion/Tijuca National Park. 1999. **Proposal for a Participatory Management Model for Tijuca National Park**. Rio de Janeiro, Report.

ISER-Institute for the Study of Religion/Tijuca National Park. 2000. **Proposal for a Participatory Management Model for Tijuca National Park**. Rio de Janeiro, Final Report.

IUCN. 1996. **IUCN RedList ofthreatenedanimals**.Gland, Switzerland. CHICO MENDES INSTITUTE FOR BIODIVERSITY CONSERVATION/ ICMBIO.**UNIDADES DE CONSERVAÇÃO**. ACCESSED IN APRIL 2013

AVAILABLE AT:
<HTTP://WWW.ICMBIO.GOV.BR/PORTAL/IMAGES/STORIES/IMGS-UNIDADES-COSERVACAO/4.%20INTRODU%C3%A7%C3%A3O.PDF>. 2013

JECUPÉ, KakaWerá. **The land of a thousand peoples**: Brazilian indigenous history told by an Indian. Editora Peirópolis, 1998.

BRAZILIAN INSTITUTE OF GEOGRAPHY AND STATISTICS - IBGE. Accessed Apr. 2013. **National basic sanitation survey**. Available at   :
<http://www.ibge.gov.br/home/estatistica/populacao/condicaodevida/pnsb/>. Rio de Janeiro: 200.

IPEA / PCRJ / PNUD, MUNICIPAL URBAN PLANNING SECRETARIAT. **Rio de Janeiro Human Development Report Environment and Sustainability**. Rio de Janeiro. Accessed on: Apr. 2013. Available at: <http://www.armazemdados.rio.-rj.gov.br PCRJ, 21 p.

KAPOS, V.; WANDELLI, E.; CAMARGO, J. L. & GANADE, G. **Edge-related changes in environment and plant responses due to forest fragmentation in central Amazonia**. In: LAURANCE, W. F. &BIERREGAARD-Jr, R. O. (eds.). Tropical Forest Remnants: ecology, management, and conservation of fragmented communities. Chicago University Press.Chicago, p. 33-44., 1997.

LEAL, MUI. **Vale do Encantado**. Accessed on: 01 April. Available at: <http://www.rioquepassou.com.br/2011/10/03/furnas-de-agassiz-anos-   50/2011>. 2003

LINO, C. F. **Mata Atlântica Biosphere Reserve**. Campinas, SP, 101p.1992.

LAURANCE, W. F. & BIERREGARD, R. O. (Eds.).**Tropical forest remnants:** Ecology,Management, and Conservation of fragmentedcommunities.The University of Chicago Press.Chicago. USA. 616 p., 1997

LAURENCE, W. F. & YENSEN,E. **Predicting the impacts of edge effects in fragmented habitats.**BiologicalConservation.1991.

LAURANCE, G; ANADRADE, A; RIBEIRO, J. E.L.S.; GIRALDO, J.P.; LOVEJOY; T.E., CONDIT, R., CHAVE, J., HARMS, E.K and D'ANGELO, S. **Rapid decay of treecommunity composition,** 2006.

MENDONÇA, H.; NUMAN, G. W.; MORAES Jr., D. F.; MARTINS, L. N.; OLIVEIRA, J. A.; SANTOS, A. C. **Exotic Fish Introduced into the State of Rio de Janeiro.** Abstracts XVI Brazilian Congress of Zoology, João Pessoa, 114p., 2005.

MOTTA, A.C.S. **The sanitary sewage network in Guanabara grows.** In: Revista de Engenharia do Estado da Guanabara, Organ of the Secretariat of Public Works, n. **1,** Estado da Guanabara: jan./mar. p. 41-58. 1965.

MINISTRY OF THE ENVIRONMENT, SECRETARIAT FOR BIODIVERSITY AND FORESTS. **Invasive alien species:** the Brazilian situation. Brasília: MMA, 2006.

ODUM, E.P. **Ecologia**. Rio de Janeiro: Guanabara, 1988.

PEIXOTO, G. L.; MARTINS, S. V.; SILVA, A. F.; SILVA, E. **Floristic Composition of the Arboreal Component of a Stretch of Atlantic Forest in the Serra da Capoeira Grande Environmental Protection Area.** Rio de Janeiro. In: Acta Botanica Brasilica, vol. 18, pp. 151-160, 2004

Management Plan - PARNA-Tijuca 2008/ IBAMA.

Tijuca National Park Management Plan - Part 2: Analysis of the Conservation Unit Region, 2008.

PESSOA, A. D. & ALMEIDA T. C. **The Rio Carioca in the city of Rio de Janeiro,** Brazil: what to preserve from its history. 2005. FIOCRUZ/ENSP.

Queiroz Fernando Fonseca de. **Brasil: Estado laico e a inconstitucionalidade da existência de símbolos religiosos em prédios públicos.**Jus Navigandi. 2005

REVISTA RAZÃO SOCIAL, O Globo, Rio de Janeiro, 15 Sep. 2009.

SECRETARY OF STATE FOR THE ENVIRONMENT - SEA. **Projects and Programmes,** 2012.

SECRETARY OF STATE FOR THE ENVIRONMENT. BIODIVERSITY AND

NATURAL RESOURCES COORDINATOR. **Riparian Forest Notebooks Riparian Forest Recovery Project Coordination Unit**. Accessed on: 01 April. Available at: <http://www.ambiente.sp.gov.br/>. São Paulo: SMA, 2009.

SILVA,Katyucha Von Kossel de. **Environmental analyst at PARNA- Tijuca, interviewed by the author by e-mail,** 2013.

SIMÕES, Manoel Ricardo. **The Shattered City:** Economic Restructuring and Municipal Emancipations in the Baixada Fluminense. Mesquita: Entorno. 2007. 298 p.:Il.

REVISTA BIO. **Brazilian Journal of Sanitation and the Environment.** Rio de Janeiro: ABES. Year XI, n. 23, July/September 2002.

ROCHA, C. F. D.; BERGALLO, H. G.; ALVES, M. A S.; VAN SLUYS, M. A **Biodiversidade nos Grandes Remanescentes Florestais do Estado do Rio de Janeiro e nas Restingas da Mata Atlântica.**RiMa. São Carlos-SP, 2003.

SAHTOURIS, Elizabeth. **From Chaos to Cosmos.** São Paulo: Interação, 1991.

TURNER, M.G. **Landscape Ecology:** effect of pattern on process. In: Annual *Review of Ecological Systems,* vol. 10, n° 3, p. 171-197, 1989.

TEIXEIRA, I.S.N.; MACHADO, D.M.C.;CASTRO, A.R.S. F; FARIAS, L.F. **A tool for understanding the appropriation of geological heritage by society:** a study of Corcovado Hill/Rio de Janeiro. *Anu. Inst. Geocienc.* [online]. 2012, vol.35, n.1, pp. 123-132.ISSN 0101-9759. http://dx.doi.org/10.11137/2012_1_123_132.

PÁDUA, J. A. A **Breath of Destruction**. *Political Thought and Environmental Criticism in Slavery Brazil (1786-1888).* Jorge Zahar Editor. Rio de Janeiro, 2002.

PRIMACK, Richard B. **A primer of conservation biology**. Sunderland, Sinauer Associates Inc., 1995, p. 66.

SILVA MATOS, D. M; SANTOS, C. J. F; CHEVALIER, D. R. **Fire and Restoration of the Largest Urban Forest of the World in Rio de Janeiro City, Brazil.**UrbanEcosystems, vol. 6(3), pp. 151-161., 2002.

TABARELLI, M.; SILVA, J. M. C.; GASCON, C. **Forest Fragmentation, Synergisms and the Impoverishment of Neotropical Forests.**Biodiversity and Conservation, vol. 13, pp. 1419-1425, 2004.

TELLES, P.C.S. **História da Engenharia no** Brasil - **século XVI a XIX.** 2. ed. Rio de Janeiro: Livros Técnicos e Científicos, 1984.

ROSAS, R. **Soil Formation in a Forested Environment, Tijuca Massif,** *RJ.* Master's Thesis, Postgraduate Programme in Geography, UFRJ. 103p., 1991.

ROWELL, A. & MOORE, P. F. **Global Review of Forest Fires.** IUCN, The World

Conservation Union. 2000.

ENCHANTED VALLEY. **SUSTAINABLE TOURISM. ALTO DA BOA VISTA.** Available at: <HTTP://WWW.VALEENCANTADO.ORG.BR/SITE/O-VALE-ENCANTADO/>. Accessed in March 2013. 2011

SEE RIO. **House approves new Forest Code. Report in:** *Veja*, Rio de Janeiro, 25 Apr. 2012.

VIANNA, M.S.R. Salubridade Domiciliar: **Uma discussão sobre sanamento básico nas favelas do município do Rio de Janeiro.**Dissertação (Mestrado em Saúde Pública)-ENSP, Fiocruz, Rio de Janeiro, 1991.

yes
I want morebooks!

Buy your books fast and straightforward online - at one of world's fastest growing online book stores! Environmentally sound due to Print-on-Demand technologies.

Buy your books online at
www.morebooks.shop

Kaufen Sie Ihre Bücher schnell und unkompliziert online – auf einer der am schnellsten wachsenden Buchhandelsplattformen weltweit! Dank Print-On-Demand umwelt- und ressourcenschonend produziert.

Bücher schneller online kaufen
www.morebooks.shop

info@omniscriptum.com
www.omniscriptum.com

OMNIScriptum

Printed by Books on Demand GmbH, Norderstedt / Germany